HARCOURT

Math

Practice Workbook

Grade 4

Harcourt

Orlando   Austin   Chicago   New York   Toronto   London   San Diego

Visit *The Learning Site!*
www.harcourtschool.com

ISBN 0-15-336476-9

21 22 23 24 -073- 11 10 09

CONTENTS

Understand Place Value

Write the value of the digit 3 in each number.

1. 4,389   **2.** 3,270   **3.** 56,223   **4.** 78,530

_____   _____   _____   _____

Compare the digits to find the value of the change.

5. 67,335 to 47,335   **6.** 45,289 to 45,889   **7.** 48,367 to 42,367

_____   _____   _____

Change the value of the number by the given amount.

8. 2,305 decreased by 200   **9.** 72,358 increased by 6,000

_____   _____

10. 46,883 decreased by 40   **11.** 29,402 increased by 40,000

_____   _____

Complete.

12. $56,891 = 50,000 + $ _____ $ + 800 + 90 + 1$

13. _____ $6,408 = 80,000 + 6,000 + 400 + 8$

14. $37,905 = $ _____ $ + 7,000 + 900 + 5$

Mixed Review

15.	**16.**	**17.**	**18.**	**19.**
420	818	77	213	633
307	128	20	501	409
+ 21	+ 66	+ 18	+ 190	+ 7

20.	**21.**	**22.**	**23.**	**24.**
100	87	98	53	110
− 22	− 24	− 69	− 8	− 56

© Harcourt

Place Value Through Hundred Thousands

Vocabulary

Write the correct letter that describes each number.

1. 340,548 _____

2. 300,000 + 40,000 + 500 + 40 + 8 _____

3. three hundred forty thousand, five
   hundred forty-eight _____

a. expanded form

b. word form

c. standard form

Write each number in two other forms.

4. 408,377

_____

_____

5. 20,000 + 600 + 30 + 2

_____

_____

6. six hundred fourteen thousand,
   two hundred thirty-nine

_____

_____

7. 892,200

_____

_____

Complete.

8. 35,309 = thirty-five _____, three hundred _____ = 30,000 +

   _____ + 300 + 9

9. 60,000 + 4,000 + _____ + 20 + 5 = _____ thousand, eight

   hundred twenty-five = _____ 4,8_____ 5

Write the value of the underlined digit.

10. 569,3<u>9</u>4 _____

11. 495,2<u>9</u>4 _____

12. <u>3</u>84,294 _____

Mixed Review

13. 39,338 − _____ = 34,338

14. 36 + 88 = _____

15. 28 − _____ = 19

Place Value Through Millions

Vocabulary

1. The period to the left of *thousands* is _____.

Write the value of the bold digit.

2. 45,**5**95,445   3. **3**,502,305   4. **7**35,495,305

_____   _____   _____

Write each number in word form.

5. 6,393,203   6. 492,203,200

_____   _____

_____   _____

_____   _____

Use place value to find each missing number.

7. 32,615,394; 32,715,394;   8. 5,398,394; 6,398,394;

_____; 32,915,394   _____; 8,398,394

9. Write the standard form of the number which is 1,000,000 less than forty-five million, three hundred twelve thousand, eight hundred.

10. Write 312,393,205 in expanded form.

_____

_____

_____

_____

Mixed Review

Complete.

11. 70,000 + 8,000 + 40 + 9 =   12. 100,000 + 60,000 + 900 + 3 =

_____   _____

13. 690 − _____ = 422   14. _____ + 222 = 879

Benchmark Numbers

Vocabulary

Fill in the blank.

1. A _____ is a known number of things that helps you understand the size or amount of a different number of things.

Use the benchmark to decide which is the more reasonable number.

2. Pennies in the jar

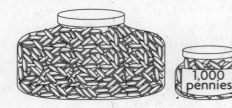

500 or 5,000

_____

3. Houses in the neighborhood

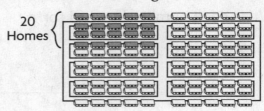

100 or 1,000

_____

4. Height of a shrub

8 in. tall

20 inches or 200 inches

_____

5. Books on a shelf

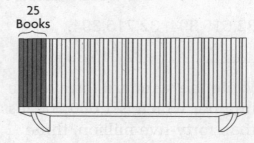

25 Books

200 or 2,000

_____

Mixed Review

6. $\begin{array}{r} 3 \\ + 8 \\ \hline \end{array}$
7. $\begin{array}{r} 9 \\ + 5 \\ \hline \end{array}$
8. $\begin{array}{r} 16 \\ + 12 \\ \hline \end{array}$
9. $\begin{array}{r} 24 \\ + 51 \\ \hline \end{array}$
10. $\begin{array}{r} 45 \\ + 22 \\ \hline \end{array}$

11. $\begin{array}{r} 31 \\ + 18 \\ \hline \end{array}$
12. $\begin{array}{r} 44 \\ + 29 \\ \hline \end{array}$
13. $\begin{array}{r} 35 \\ - 17 \\ \hline \end{array}$
14. $\begin{array}{r} 35 \\ - 27 \\ \hline \end{array}$
15. $\begin{array}{r} 59 \\ - 31 \\ \hline \end{array}$

16. $12 + 11$ _____

17. $19 + 49$ _____

18. $62 + 21$ _____

Problem Solving Skill

Use a Graph

The United States Department of Agriculture has named 5 food groups and recommends a maximum number of daily servings from each group.

Maximum Daily Servings	
dairy	◎ ⦅
meat	◎ ⦅
vegetables	◎ ◎ ⦅
fruit	◎ ◎
bread and cereal	◎ ◎ ◎ ◎ ◎ ⦅

Key: Each ◎ stands for 2 servings.

For 1–8, use the graph.

1. What is the maximum recommended number of meat servings?

   _____

2. Which two food groups have the same number of recommended servings?

   _____

3. Of which food groups can you eat more than four servings per day?

   _____

4. Of which food group can you eat the most servings?

   _____

5. Today, Erika ate 5 servings of meat. How would you represent this on the pictograph?

   _____

6. What is the total number of fruit and vegetable servings recommended?

   _____

7. Rolanda has eaten 7 servings from the bread and cereal group today. How many more servings could she have?

   _____

8. At breakfast, Jamika's banana counted as 2 fruit servings. How many more fruit servings could she have today?

   _____

Mixed Review

What is the value of the digit 7?

9. 1,762   10. 7,900,631   11. 44,072,461   12. 817,535

_____

Compare Numbers

Compare using the number line. Write the greater number.

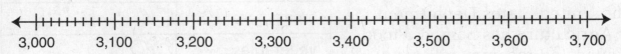

3,000   3,100   3,200   3,300   3,400   3,500   3,600   3,700

1. 3,568 or 3,658   **2.** 3,102 or 3,020   **3.** 3,460 or 3,500

_____   _____   _____

4. 3,215 or 3,196   **5.** 3,609 or 3,099   **6.** 3,254 or 3,520

_____   _____   _____

Compare. Write <, >, or = in each ◯.

7. 3,489 ◯ 3,578   **8.** 25,899 ◯ 25,890   **9.** 75,673 ◯ 75,673

10. 3,142,355 ◯ 314,235   **11.** 33,452,236 ◯ 35,235,032

Find all of the digits that can replace each ☐.

12. 6☐7,348 < 647,348   **13.** 35,468,245 < 35,468,☐45

_____   _____

Mixed Review

14. Write 8,000,000 + 30,000 +
5,000 + 400 + 30 + 2 in
standard form.

15. Write 32,883 in word form.

_____

_____

16. What digit is in the ten
thousands place in 32,456,922?

17. Write the value of the digit 8 in
the number 385,722.

_____

_____

18. Round 7,899 to the nearest
hundred.

19. Round 42,616 to the nearest ten.

_____

_____

Order Numbers

Write the numbers in order from *least* to *greatest*.

1. 15,867; 15,394; 15,948; 15,493

2. 65,447; 65,743; 65,446, 65,395

_____

3. 249,330; 247,449; 248,390

4. 3,456,490; 3,458,395; 3,359,498

_____

Write the numbers in order from *greatest* to *least*.

5. 45,387; 48,339; 47,110

6. 252,484; 259,793; 258,932

_____

7. 2,783,859; 2,788,394; 2,937,383

8. 360,839; 45,395; 366,395

_____

9. 4,671,302; 4,716,230; 4,716,200

10. 740,516; 74,506; 740,605

_____

Write all of the digits that can replace each □.

11. 4,599 < 4,63□ < 4,634

12. 3,554,684 > 3,□69,304 >
3,184,394

_____

Mixed Review

13.   25
    + 42

14.   99
      21
    + 86

15.   95¢
    − 43¢

16.   78¢
    − 24¢

17.   13
      74
    + 26

18. Stacey jogged for 25 minutes on Saturday and 38 minutes on Tuesday. How much longer did she jog on Tuesday than on Saturday?

19. Rolanda completed 12 home-work problems before dinner and 18 after dinner. How many homework problems did she complete altogether?

_____

© Harcourt

Name _____

Problem Solving Strategy

Make a Table

Make a table to solve.

The Sahara Desert in Africa has an area of about 3,500,000 square miles. The Simpson Desert in Australia has an area of about 56,000 square miles.

Desert	Continent	Approximate Area (sq mi)

In North America, the Mojave Desert has an area of about 15,000 square miles; and the Kalahari Desert in Africa has an area of about 225,000 square miles.

1. Which desert has the greatest area?

_____

2. Which two deserts are located on the same continent?

_____

3. Which deserts have an area of less than 100,000 square miles?

_____

4. On which continent is the desert with the least area located?

_____

Mixed Review

5. Write $3,000,000 + 20,000 + 5,000 + 300 + 70 + 2$ in standard form.

_____

6. Write the numbers in order from *least* to *greatest*: 254,879; 2,254,920; 1,678,305; 353,502.

_____

Compare. Write $<$, $>$, or $=$ in the $\bigcirc$.

7. $354{,}992 \bigcirc 288{,}492$

8. $7{,}394{,}398 \bigcirc 7{,}394{,}398$

9. $394{,}234 \bigcirc 3{,}294{,}394$

10. $6{,}187{,}249 \bigcirc 61{,}872{,}490$

11. $\begin{array}{r} 9{,}421{,}720 \\ -\ 6{,}198{,}135 \\ \hline \end{array}$

12. $\begin{array}{r} 210{,}076 \\ +\ 935{,}811 \\ \hline \end{array}$

13. $\begin{array}{r} 8{,}176{,}553 \\ +\ \ \ \ 30{,}602 \\ \hline \end{array}$

14. $\begin{array}{r} 172{,}442 \\ -\ 172{,}435 \\ \hline \end{array}$

15. $786 - 421 =$ _____

16. $2{,}779 - 460 =$ _____

PW8 Practice

Round Numbers

Round each number to the nearest thousand.

1. 5,339

2. 9,895

3. 75,367

4. 22,022

_____ _____ _____ _____

5. 5,600,679

6. 1,354,029

7. 283,966

8. 636,592

_____ _____ _____ _____

Round each number to the place of the underlined digit.

9. 6,333

10. 837

11. 8,021

_____ _____ _____

12. 45,935

13. 356,882

14. 502,446

_____ _____ _____

15. 24,546

16. 888,044

17. 47,164

_____ _____ _____

18. 1,999,444

19. 1,366,901

20. 9,203,774

_____ _____ _____

Mixed Review

21. $9 + 4 + 5 =$ _____

22. $27 + 33 + 59 =$ _____

23. $48 - 29 =$ _____

24. $\begin{array}{r} 6 \\ \times\ 2 \\ \hline \end{array}$

25. $\begin{array}{r} 8 \\ \times\ 5 \\ \hline \end{array}$

26. $\begin{array}{r} 9 \\ \times\ 8 \\ \hline \end{array}$

27. $\begin{array}{r} 7 \\ \times\ 7 \\ \hline \end{array}$

28. What is the value of the digit 7 in 478,394?

29. What is the value of the digit 5 in 5,394,332?

_____ _____

Use Mental Math Strategies

For 1–4, use the *break apart* strategy.

1. 49 + 16   **2.** 73 − 43   **3.** 46 − 12   **4.** 91 − 63

_____   _____   _____   _____

For 5–8, use the *make a ten* strategy.

5. 94 − 56   **6.** 88 + 31   **7.** 72 − 39   **8.** 84 + 46

_____   _____   _____   _____

For 9–28, add or subtract mentally. Tell the strategy you used.

9. 78 + 46   **10.** 61 − 16   **11.** 40 + 24   **12.** 37 − 19

_____   _____   _____   _____

13. 64 − 28   **14.** 45 + 48   **15.** 58 + 32   **16.** 67 + 43

_____   _____   _____   _____

17. 82 − 53   **18.** 66 − 27   **19.** 53 − 23   **20.** 75 + 61

_____   _____   _____   _____

21. 51 + 38   **22.** 49 + 21   **23.** 82 − 46   **24.** 49 − 31

_____   _____   _____   _____

25. 83 + 72   **26.** 28 − 19   **27.** 93 − 38   **28.** 26 + 23

_____   _____   _____   _____

Mixed Review

Round each number to the place given.

29. 568,303; ten thousand   **30.** 35,405,203; million   **31.** 596,305,003; ten million

_____   _____   _____

Write the numbers in order from *least* to *greatest*.

32. 568,394; 395,205; 562,304 _____

33. 458,404,305; 451,402,305; 455,305,203 _____

_____

Name _____

Estimate Sums and Differences

Estimate the sum or difference.

1. 7,379
    + 5,496

2. $479,150
    − $371,271

3. 612,797
    + 811,035

4. 638,113
    − 415,327

5. 5,324
    + 2,468

6. $6,372
    − $4,047

7. 721,379
    +   15,496

8. $3,016
    − $2,849

9. 8,492
    + 1,346

10. 846,134
     − 794,134

11. 461,137
     + 91,214

12. $9,263
     − $ 489

Write the missing digit for the estimated sum or difference.

13. ☐46,164
     − 471,467
     ─────────
     100,000

14. 23,497
     + ☐2,464
     ─────────
     80,000

15. 631,431
     − ☐6,497
     ─────────
     520,000

16. ☐79,431
     + 231,587
     ─────────
     400,000

17. ☐21,863
     − 135,632
     ─────────
     300,000

18. 54,961
     + ☐5,246
     ─────────
     70,000

19. ☐45,239
     − 32,878
     ─────────
     170,000

20. 58,138
     + ☐3,245
     ─────────
     90,000

Mixed Review

21. 27
     + 49

22. 31
     + 64

23. 92
     + 11

24. 87
     + 34

25. 16
     + 77

Add and Subtract to 4-Digit Numbers

Find the sum or difference. Estimate to check.

1. $\begin{array}{r} 7,503 \\ -\ 3,598 \\ \hline \end{array}$

2. $\begin{array}{r} 2,178 \\ +\ 3,703 \\ \hline \end{array}$

3. $\begin{array}{r} 5,527 \\ 2,978 \\ +\ 1,852 \\ \hline \end{array}$

4. $\begin{array}{r} 3,092 \\ 1,574 \\ +\ 1,296 \\ \hline \end{array}$

5. $\begin{array}{r} 1,468 \\ +\ 1,090 \\ \hline \end{array}$

6. $\begin{array}{r} 2,714 \\ -\ 1,833 \\ \hline \end{array}$

7. $\begin{array}{r} 2,131 \\ 1,574 \\ +\ 1,078 \\ \hline \end{array}$

8. $\begin{array}{r} 2,858 \\ +\ 1,670 \\ \hline \end{array}$

9. $4,375 + 5,839$

10. $4,793 + 2,988 + 8,349$

11. $5,707 - 2,596$

12. $3,872 + 2,396 + 7,236$

For 13–20, find the missing digit.

13. $\begin{array}{r} 7,13\square \\ -\ 2,467 \\ \hline 4,671 \end{array}$

14. $\begin{array}{r} 4,135 \\ +\ \square,252 \\ \hline 5,387 \end{array}$

15. $\begin{array}{r} 5,6\square7 \\ -\ 3,684 \\ \hline 1,953 \end{array}$

16. $\begin{array}{r} 6,465 \\ +\ 1,\square68 \\ \hline 8,233 \end{array}$

17. $\begin{array}{r} 5,\square23 \\ +\ 1,820 \\ \hline 7,043 \end{array}$

18. $\begin{array}{r} 9,465 \\ -\ 8,4\square7 \\ \hline 968 \end{array}$

19. $\begin{array}{r} \square,254 \\ +\ 2,849 \\ \hline 7,103 \end{array}$

20. $\begin{array}{r} 6,102 \\ -\ 4,58\square \\ \hline 1,517 \end{array}$

Mixed Review

21. $10 + 10 + 10 + 10 =$ _____

22. $5 + 5 + 5 + 5 + 5 =$ _____

23. $42 - 21 =$ _____

24. $63 - 12 =$ _____

Subtract Across Zeros

Find the difference. Estimate to check.

1. 3,000
   − 2,780

2. 4,003
   − 2,232

3. 8,005
   − 5,004

4. 6,200
   − 4,816

5. 5,700
   − 1,751

6. 9,100
   − 3,759

7. 20,000
   − 13,652

8. 10,000
   − 2,842

9. 90,000
   − 66,536

10. 50,000
    − 13,747

11. 20,000
    − 15,136

12. 50,075
    − 32,097

13. 70,000
    − 29,134

14. 50,000
    − 19,673

15. 70,006
    − 43,989

16. 20,000
    − 9,342

Compare. Write <, >, or = in each ◯.

17. 2,006 − 1,513 ◯ 4,075 − 3,209

18. 7,004 − 6,315 ◯ 5,075 − 4,897

19. 8,003 − 3,695 ◯ 7,473 − 2,127

20. 9,200 − 5,861 ◯ 6,153 − 2,814

21. 3,009 − 1,819 ◯ 8,006 − 6,952

22. 4,284 − 2,651 ◯ 9,000 − 7,367

Mixed Review

23. 6,491
    + 8,034

24. 9,403
    + 199

25. 8,662
    + 8,449

26. 7,361
    + 9,170

27. 2,649
    + 3,427

28. 2,831
    + 6,923

29. 1,424
    + 3,462

30. $2,455
    +$3,119

Choose a Method

Find the sum or difference. Estimate to check.

1. 213,742
    + 170,045

2. 408,587
    − 345,128

3. 248,232
    + 236,816

4. 684,004
    − 195,751

5. 661,119
    − 423,384

6. 358,379
    + 264,175

7. 568,075
    − 372,097

8. 468,951
    + 236,175

Compare. Write <, >, or = in each ○.

9. 561,257 − 346,052 ○ 846,735 − 612,435

10. 257,132 + 153,087 ○ 210,735 + 128,307

11. 976,034 − 780,347 ○ 461,597 − 265,910

Find the missing digit.

12.   4□6,341
    − 275,132
       221,209

13.   682,318
    − 248,1□6
       434,142

14.   945,132
    + 153,□02
     1,098,734

Mixed Review

Estimate the sum or difference.

15.   6,842
    + 2,981

16.   1,132
      2,074
    + 2,596

17.   4,008
    − 2,567

18. 6,921 − 4,071 = _____

19. 3,460 − 782 = _____

20. 8,130 − 3,471 = _____

21. 1,197 − 238 = _____

Problem Solving Skill

Estimate or Find Exact Answers

Tell whether an estimate or an exact answer is needed. Solve.

Item	Price
Poster	$5.95
Souvenir Cup	$3.50
Animal Encyclopedia	$10.00
Hat	$7.50
T-shirt	$12.00

1. Mitchell bought a hat and a poster. How much change will he get from $20.00?

_____

2. About how much money does someone need to buy one of each item?

_____

3. Tracy wants to buy a t-shirt and a souvenir cup. If she has $15.00, does she have enough? Explain your answer.

_____

4. Maurice had $15.00. He bought a hat. About how much money is left? Is it enough to buy a poster?

_____

5. Tanisha and Shauna want to share the Animal Encyclopedia. Tanisha has $4.75 and Shauna has $3.25. How much more money do they need to buy the book?

_____

6. D'Angelo wants to buy lunch for $5.75 and buy a poster and souvenir cup. About how much money should he bring to the zoo?

_____

Mixed Review

7.
$1.73
+$0.14

8.
$10.00
−$ 8.59

9.
6,285
− 3,119

10.
16,212
+ 42,080

11.
$19.27
+$11.27

12.
$3,204
−$2,413

13.
5,320
+1,375

14.
9,862
−7,361

15.
$3,228
+$4,228

16.
40,000
− 8,613

© Harcourt

Name _____

Expressions

Vocabulary

Complete each sentence.

1. An _____ has numbers
   and operation signs but no equal sign.

2. A _____ is a letter or
   symbol that represents any number you don't know.

Write an expression.

3. There are 16 people at the club
   meeting, and then 9 more come.

4. Rob had 23 baseball cards. He
   found 4 more cards then gave
   9 of them away.

_____

Write an expression. Choose a variable for the unknown.

5. Karen had 30 problems to do for
   homework. She did some before
   dinner and 21 after dinner.

6. Adam had 23 baseball cards. He
   found 4 more cards, then gave
   some of them away.

_____   _____

Find the value of each expression.

7. $25 - (6 + 11)$ 8. $43 - 9 - 8$ 9. $(56 - 32) + 4$

_____   _____   _____

Find the value of each expression if $x = 14$ and $y = 8$.

10. $40 - x$ 11. $(y + 12) + x$ 12. $52 - (y + 9)$

_____   _____   _____

Mixed Review

13. $\begin{array}{r} 2{,}112 \\ + 5{,}899 \\ \hline \end{array}$ 14. $\begin{array}{r} 85{,}584 \\ - 29{,}920 \\ \hline \end{array}$ 15. $\begin{array}{r} 50{,}008 \\ - 28{,}251 \\ \hline \end{array}$ 16. $\begin{array}{r} 3{,}804 \\ + 9{,}156 \\ \hline \end{array}$

Addition Properties

Vocabulary

Complete the sentences.

1. The _____ Property of
   Addition states that the way addends are grouped does
   not change the sum.

2. The _____ Property of
   Addition states that the numbers can be added in any
   order and the sum will be the same.

_____

Find the missing number, and tell which addition property you used.

3. $5 + (2 + 1) = (5 + \blacksquare) + 1$ 4. $72 + 18 = \blacksquare + 72$ 5. $\blacksquare + 0 = 24$

_____   _____   _____

6. $45 + 9 = 9 + \blacksquare$ 7. $15 + \blacksquare = 15$ 8. $(4 + \blacksquare) = (5 + 4)$

_____   _____   _____

Change the order or group the addends so that you can add them mentally.
Find the sum. Tell which property you used.

9. $22 + 14 + 8$ 10. $(12 + 17) + 3$ 11. $20 + 16 + 80$

_____   _____   _____

12. $4 + 2 + 106$ 13. $25 + (15 + 28)$ 14. $(145 + 290) + 110$

_____   _____   _____

Mixed Review

What is the value of the digit 4 in each number?

15. $84,250,000$ 16. $1,024$ 17. $340,112$

_____   _____   _____

© Harcourt

Equations

Vocabulary

Complete the sentence.

1. An _____ is a number sentence stating that two amounts are equal.

Write an equation for each. Choose a variable for the unknown.

2. There are 26 students in Mrs. Philips' class. Fifteen are boys. The rest are girls.

   _____

   _____

3. Arturo has 4 posters. He buys some more posters. Now he has 12 posters.

   _____

   _____

4. Mr. Tran has 45 students in gym class. Thirty-two are playing volleyball. The rest are jogging.

   _____

   _____

5. Christine adds 4 coins to her piggy bank. There are now 83 coins.

   _____

   _____

Solve the equation by using mental math. Check your solution.

6. $x + 5 = 12$

   _____

7. $9 + y = 13$

   _____

8. $w - 8 = 4$

   _____

9. $18 - c = 15$

   _____

10. $5 = 9 - a$

   _____

11. $b - 4 = 3$

   _____

Mixed Review

Find the value.

12. $3 + (20 - 12)$

   _____

13. $(5 + 8) - (2 + 7)$

   _____

14. $25 - (4 + 6)$

   _____

Patterns: Find a Rule

Find a rule. Write the rule as an equation. Use the equation to extend the pattern.

1.

Input	Output
x	**y**
6	12
14	■
9	15
11	17

2.

Input	Output
a	**b**
18	10
9	1
12	4
15	■

3.

Input	Output
r	**k**
45	39
27	■
18	12
21	15

4.

Input	Output
t	**m**
13	■
8	20
17	29
3	15

_____   _____   _____   _____

_____   _____   _____   _____

Use the rule and equation to make an input/output table.

5. Add 8.
$t + 8 = p$

Input	Output

6. Subtract 3.
$w - 3 = t$

Input	Output

7. Add 14.
$c + 14 = m$

Input	Output

8. Subtract 28.
$b - 28 = g$

Input	Output

9. Add 23.
$g + 23 = y$

Input	Output

10. Subtract 32.
$m - 32 = w$

Input	Output

11. Subtract 9.
$x - 9 = b$

Input	Output

12. Add 28.
$t + 28 = r$

Input	Output

Mixed Review

Round to the nearest million.

13. 58,405,303

14. 492,920,302

15. 289,810,304

_____   _____   _____

Balance Equations

Tell whether the equations are true. Write *yes* or *no*. Explain.

1. 1 quarter = 2 dimes

_____

2. 1 dime − 2 pennies =
1 nickel + 3 pennies

_____

3. 3 dimes + 2 nickels =
40 pennies

_____

4. 4 pennies + 1 quarter =
3 dimes

_____

5. 2 quarters =
3 dimes + 2 nickels

_____

6. 3 dimes + 3 nickels + 1 quarter =
1 quarter + 2 dimes + 1 nickel

_____

Complete to make the equation true.

7. 3 dimes + 1 nickel + 2 pennies =
1 quarter + 1 dime + ___?___

_____

8. 3 quarters + 1 nickel =
2 quarters + ___?___

_____

9. $19 + 3 = $ ___?___ $+ 19$

_____

10. $12 + 4 = 6 + $ ___?___

_____

11. $2 + 7 + $ ___?___ $= 10 + 7$

_____

12. $15 + 6 = 7 + 7 + $ ___?___

_____

13. $22 + 8 + 1 = 25 + $ ___?___

_____

14. ___?___ $+ 5 = 10 + 3 + 1$

_____

Mixed Review

Find the value of each expression.

15. $(9 + 11) − (4 + 4)$ _____

16. $72 − (41 + 6)$ _____

17. $(49¢ − 22¢) + 17¢$ _____

18. $(15 + 11) − 6$ _____

19. $(43 − 8) + (7 − 5)$ _____

20. $(90 − 21) + 17$ _____

Problem Solving Strategy

Act It Out

Act it out to solve.

There is a contest among the different grades at Memorial Elementary school. The contest lasts for two weeks. The first grade to collect 20 bags of recyclables wins a pizza party.

1. Students from Grade 2 brought in 4 bags then brought in 7 more bags. How many more bags do they need to win?

_____

2. At the end of the contest, Grade 5 had collected 16 bags. If they collected 5 bags in Week 2, how many did they collect in Week 1?

_____

3. Grades 1 and 3 have decided to work together. If Grade 1 brought in 12 bags and Grade 3 brought in 16 bags, how many do they have altogether?

_____

4. If Grade 6 collects 9 bags in Week 1 and 8 bags in Week 2, how many more do they have than Grade 2?

_____

5. At the end of the contest, Grade 4 had collected 5 more bags than Grades 1 and 3 combined. How many bags of recyclables did Grade 4 collect?

_____

6. How many more bags should Grade 2 collect so that they have the same number as Grades 1 and 3 combined?

_____

Mixed Review

Use the rule and equation to complete the input/output table.

7. Add 6.
$x + 6 = z$

Input	Output

8. Subtract 31.
$m - 31 = r$

Input	Output

9. Add 19.
$p + 19 = s$

Input	Output

10. Subtract 25.
$c - 25 = a$

Input	Output

Telling Time

Write the time as shown on a digital clock.

1. 7 minutes after 3

2. 28 minutes before 11

3. 15 minutes after 5

_____

_____

_____

4. 18 minutes after 2

5. 3 minutes after 12

6. 15 minutes before 7

_____

_____

_____

Write the time shown on the clock in 2 different ways.

7.

8.

9.

_____   _____   _____

_____   _____   _____

Write the letter of the unit used to measure the time.

10. to take a shower   _____   **a.** days

11. to drive across the United States   _____   **b.** hours

12. to button a button   _____   **c.** minutes

13. to get a night's sleep   _____   **d.** seconds

Mixed Review

14. Find the value of the expression.
$59 - (32 + 12)$ _____

15. Find the value of the expression.
$(28 - 9) - (4 + 8)$ _____

16. Order from least to greatest:
37,623; 37,326; 36,723

17. Estimate the difference between 47,791 and 35,167.

_____   _____

Elapsed Time

Vocabulary

Complete the sentence.

1. _____ is the time that passes
   from the start to the end of an activity.

Find the elapsed time.

2. **start:** 7:30 A.M.   3. **start:** 8:05 A.M.   4. **start:** 9:12 P.M.
   end: 3:15 P.M.   **end:** 9:55 A.M.   **end:** 11:28 P.M.

_____   _____   _____

Complete the tables.

	Start Time	End Time	Elapsed Time
5.	7:20 A.M.		1 hr 30 min
6.	10:12 A.M.	4:15 P.M.	

	Start Time	End Time	Elapsed Time
7.	11:50 A.M.		2 hr 5 min
8.		6:35 P.M.	4 hr 9 min

9. Use the tour schedule to find
   how much more time the
   Red Coach Tour takes than the
   Blue Coach Tour.

_____

TOURS OF NEW YORK CITY	
Red Coach Tour	**Blue Coach Tour**
Leaves 9:45 A.M.	Leaves 1:20 P.M.
Returns 2:30 P.M.	Returns 3:55 P.M.

Mixed Review

Add or subtract. Estimate to check.

10.   5,979   11.   3,004   12.   2,753   13.   7,000
    + 2,536   − 1,536   + 1,265   − 4,264

Problem Solving Skill

Sequence Information

Mr. Anderson is taking his history class to a museum. The students will take a tour, view 2 movies, and visit the costume room. The bus will drop the class off at 9:15 A.M. and pick them up at 3:30 P.M. Lunch will be from 12:15 P.M. to 12:45 P.M. Tours of the museum last 1 hour and 15 minutes.

Revolutionary Heroes Movie	
running time: 45 min	
9:00 A.M.	1:00 P.M.
10:00 A.M.	2:00 P.M.
11:00 A.M.	3:00 P.M.

Battlegrounds Movie	
running time: 37 min	
9:30 A.M.	1:30 P.M.
10:30 A.M.	2:30 P.M.
11:30 A.M.	5:00 P.M.

1. Will the class be able to see both movies before lunch? If so, name a schedule.

   _____

   _____

   _____

2. If the class begins the museum tour at 9:40 A.M., will it be able to see *Revolutionary Heroes* and still be ready for lunch at 12:15 P.M.? Explain.

   _____

   _____

   _____

3. If the class visits the costume room at 1:45 P.M. and stays for one hour and 10 minutes, can it view *Revolutionary Heroes* and be ready to meet the bus?

   _____

   _____

   _____

4. Make a schedule for the class which includes both movies, a tour of the museum, and a visit to the costume room.

My Museum Tour Schedule	
Lunch	12:15 P.M.–12:45 P.M.

Mixed Review

5. 370,716
   − 192,408

6. 971,858
   − 863,245

7. 4,330,629
   + 6,197,550

8. 3,606,117
   − 3,432,980

© Harcourt

Name _____

Elapsed Time on a Calendar

For 1–3, use the calendars.

Camp Windy	
Session 1:	Jul 13–Jul 17
Session 2:	Jul 27–Jul 31
Session 3:	Aug 3–Aug 14

June

Sun	Mon	Tue	Wed	Thu	Fri	Sat
	1	2	3	4	5	6
7	8	9	10	11	12	13
14	15	16	17	18	19	20
21	22	23	24	25	26	27
28	29	30				

1. The camp director bought art supplies 4 weeks before the beginning of the first session of camp. On what date did she buy art supplies?

_____

July

Sun	Mon	Tue	Wed	Thu	Fri	Sat
			1	2	3	4
5	6	7	8	9	10	11
12	13	14	15	16	17	18
19	20	21	22	23	24	25
26	27	28	29	30	31	

2. In Session 3, the campers put on a puppet show on the second Wednesday of the session. What was the date of the puppet show?

_____

3. Jim plans to attend Session 2 of camp. His last day of school is June 19. About how many weeks of summer vacation will Jim have before camp begins?

_____

August

Sun	Mon	Tue	Wed	Thu	Fri	Sat
						1
2	3	4	5	6	7	8
9	10	11	12	13	14	15
16	17	18	19	20	21	22
23	24	25	26	27	28	29
30	31					

Mixed Review

Find the value of each expression.

4. $125 - (65 + 22)$

5. $234 - (24 - 13)$

6. $4,590 - (1,293 - 389)$

_____

Round to the nearest ten thousand.

7. $472,099$

8. $939,658$

9. $3,514,811$

_____

© Harcourt

Collect and Organize Data

Vocabulary

Complete the sentence.

1. The numbers in the _____ column show the sum as each new line of data is entered.

For 2–3, use the table.

FROZEN POPS SOLD		
Day	Frequency (Number of Frozen Pops)	Cumulative Frequency
Monday	15	15
Tuesday	24	39
Wednesday	19	58
Thursday	9	67
Friday	21	88

2. The cumulative frequency for Wednesday is _____. This is the sum of the numbers in the frequency column for which days?

_____, _____, and _____.

3. How many frozen pops in all were sold on Monday and Tuesday?

_____

Mixed Review

Order the numbers from *greatest* to *least*.

4. 234,358; 23,208; 23,098

5. 12,214; 342,351; 120,142

_____   _____

6. 342,253; 34,235; 34,270

7. 824,723; 8,247; 82,492

_____   _____

Find Mean, Median, and Mode

Vocabulary

Complete each sentence.

1. In a set of numbers ordered from the least to the greatest, the
   number in the middle is called the _____, and the number
   that occurs most often is called the _____. The _____
   is the number found by dividing the sum of a set of numbers by
   the number of addends.

For 2–5, use the table.

2. Find the mean of the number of
   puppies born.

   _____

PUPPIES BORN	
Litter	Puppies
1	5
2	3
3	6
4	5
5	1

3. Find the median of the number
   of puppies born.

   _____

4. Find the mode of the number of
   puppies born.

   _____

5. What if there was 1 puppy born
   in the sixth litter. Would the
   mode change? Explain.

Mixed Review

Round each number to the nearest hundred.

6. 56,298 _____

7. 355,207 _____

8. 514,899 _____

9. 29,909 _____

10. 17,893 _____

11. 99,903 _____

Find n.

12. $4 + n = 3 + 4$ _____

13. $15 + 5 = 12 + n$ _____

14. $8 + n = 10 + 6$ _____

15. $6 + 5 = 9 + n$ _____

16. $n + 7 = 19 + 12$ _____

17. $23 + n = 32 + 4$ _____

© Harcourt

Read Line Plots

Vocabulary

Complete the sentences.

1. A _____ is a graph that shows data along a number line.

2. The difference between the greatest and the least numbers

   in a set of data is called the _____.

For 3–4, use the line plot on the right.

3. The X's on this line plot represent the number of students. What do the numbers on the line plot represent?

   _____

4. What number of children do the most students have in their families?

   _____

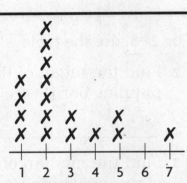

Number of Children in Family

5. Use the data in the table to complete the line plot.

Slices of Pizza Eaten at a Party						
Number of Slices	0	1	2	3	4	5
Number of Students	//	⑷ /	⑷	///	/	//

Slices of Pizza Eaten at a Party

Mixed Review

Write each number in standard form.

6. 100,000 + 50,000 + 4,000

   _____

7. ninety-six thousand

   _____

8. nine hundred seventy thousand, eight hundred fifty-two

   _____

9. 400,000 + 80 + 8

   _____

Make Stem-and-Leaf Plots

Vocabulary

Complete the sentences.

1. A _____ shows groups of data organized by place value.

2. Each tens digit is called a _____.

3. The ones digits are called the _____.

The stem-and-leaf plot below shows the scores that fourth-grade students made in a spelling contest. For 4–6, use the stem-and-leaf plot.

4. What are the least and the greatest scores?

_____

5. What is the mode of the contest scores?

_____

6. What is the median of the contest scores?

_____

Spelling Scores

Stem	Leaves
6	8 8 9 9
7	2 3 5 5 6
8	4 4 6 7 8 8 8
9	1 2 2 3 4 5 5

$6|8 = 68$

Mixed Review

Find the value of n.

7. $5 \times 6 = n$ _____   8. $9 \times 4 = n$ _____   9. $6 \times 9 = n$ _____

10. $n - 3 = 4$ _____   11. $7 + 12 = n$ _____   12. $63 \div n = 9$ _____

13. $10 + n = 13$ _____   14. $7 \times n = 56$ _____   15. $8 \times n = 64$ _____

16. Round 39,457 to the nearest 10,000.

_____

17. Ted bought eggs for $1.98, milk for $2.19, and bread for $1.10. What change should he receive from $10.00? _____

Compare Graphs

Vocabulary

Complete the sentence.

1. The _____ is the series of numbers placed at fixed distances on the side of a graph.

2. The _____ of a graph is the difference between any two numbers on the scale.

___

For 3–6, use the graph.

3. What is the interval of the scale in the graph?

    _____

4. How would the bars change in the graph if the interval were 1?

    _____

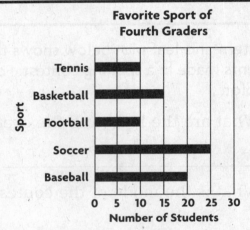

Favorite Sport of Fourth Graders

5. Describe how the bars in the graph would look if you made a new graph, using a scale interval of 10.

6. Suppose the scale of a bar graph is 0, 4, 8, 12, 16, 20. Describe the bar length that would represent the number 10.

    _____

    _____

Mixed Review

7. $55 + 23$ _____   8. $44 - 23$ _____   9. $12 + 34$ _____   10. $87 + 12$ _____

11. 5×6 _____   12. $72 \div 9$ _____   13. 12×12 _____   14. $45 \div 5$ _____

15. A baker can make 8 batches of cookies an hour. How many batches of cookies can the baker make in 7 hours?

    _____

16. Kim has a scarf. It has a red stripe, a blue stripe, then a white stripe. This pattern repeats. What color is the eighth stripe?

    _____

Problem Solving Strategy

Make a Graph

Vocabulary

Complete the sentences.

1. We can use a _____ to help see information easily.

---

2. The table shows the number and types of books students checked out of the school library last week. Make a pictograph of the data. Then find the difference between the number of mystery books checked out and the number of fantasy books checked out.

NUMBER OF BOOKS CHECKED OUT OF SCHOOL LIBRARY	
Type of Book	Number of Books
Mystery	90
Adventure	20
Fantasy	50
Nonfiction	40

Mixed Review

3. Find the median and the mode of these numbers.

14, 17, 18, 14, 12, 16. _____

4. $12.75 + $13.22 _____

5. $80.00 − $25.98 _____

Make Bar and Double-Bar Graphs

Vocabulary

Complete the sentence.

1. A _____ is used to compare similar kinds of data.

Cost of Bulbs (per package of 25)		
Type of Bulb	Kevin's Flowers	Hillside Nursery
Daffodil	$18.00	$14.00
Tulip	$10.00	$12.00
Hyacinth	$21.00	$12.00
Crocus	$5.00	$7.00

2. Make a double-bar graph to compare the cost of bulbs at Kevin's Flowers and at Hillside Nursery. Use the data from the table above. Choose an appropriate scale. Include a title, labels, a scale, and a key for both stores.

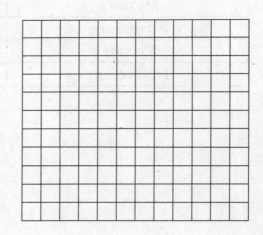

Mixed Review

3. Which is greater, 420,391 or 402,931?

4. Round 225,770 to the nearest thousand.

_____

5. Estimate. 893,232 + 281,932

6. What is the sum of 259,739 and 927,492?

_____

Read Line Graphs

Vocabulary

Complete the sentence.

1. A _____ uses a line to show how something changes over a period of time.

Joyce made this line graph to show the number of pages she read each day in a mystery book. For 2–6, use the graph.

2. On what day did Joyce read the most pages? the fewest pages?

   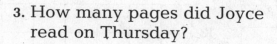
   _____

3. How many pages did Joyce read on Thursday?

   _____

4. On which two days did Joyce read the same number of pages?

   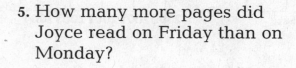
   _____

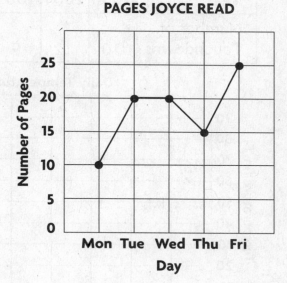

PAGES JOYCE READ

5. How many more pages did Joyce read on Friday than on Monday?

   _____

6. How many pages did Joyce read altogether from Monday through Friday?

   _____

Mixed Review

7.	35,859 + 91,847	8.	680,005 − 490,948	9.	5,940,394 − 2,518,624	10.	9,848,664 + 8,842,231
11.	762,063 − 410,978	12.	248,671 + 99,348	13.	7,100,003 − 6,471,691	14.	8,317,062 + 4,065,594

Make Line Graphs

For 1–2, complete the line graph.

1.

Daily Temperature							
Day	Sun	Mon	Tue	Wed	Thu	Fri	Sat
Temperature (in °F)	65	70	85	75	70	80	80

2.

Touchdowns Made					
Year	2000	2001	2002	2003	2004
Number of Touchdowns	10	12	9	15	18

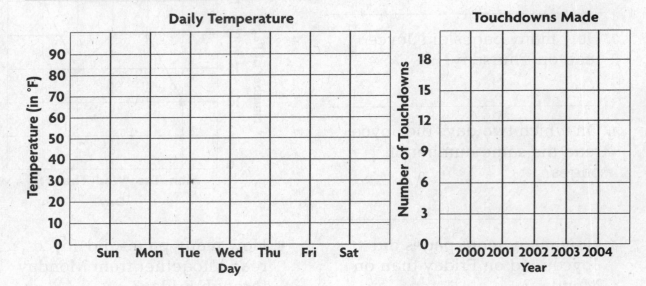

3. Which day had the highest temperature? What was the temperature on that day?

_____

4. Describe any trends in the number of touchdowns made.

_____

_____

Mixed Review

5. Compare. Use <, >, or =.
   7,458 ◯ .8,125 − 304

6. What number is 100,000 greater than 1,825,435?

_____

Name _____

Make Circle Graphs

Vocabulary

Complete the sentence.

1. A _____ shows data as a whole made up of different parts.

For 2–3, make a circle graph.

2.

Favorite Subject	
Subject	Number of Votes
Math	50
Science	20
Reading	30

3.

Calvin's Allowance	
Activity	Amount
Games	$6
Zoo	$3
Magazine	$1

FAVORITE SUBJECT

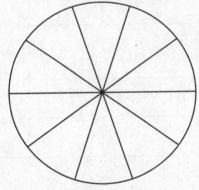

CALVIN'S ALLOWANCE

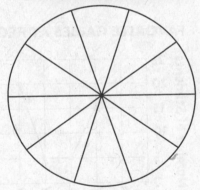

4. Chris takes care of 10 birds at the farm. Four of them are ducks, and the rest are geese. How would you show this in a circle graph?

_____

_____

Mixed Review

5. 2,793
+ 5,095

6. 20,050
− 11,903

7. 82,008
+ 5,996

8. 73,021
− 9,093

Choose an Appropriate Graph

For 1–4, write the kind of graph or plot you would choose.

1. to show a record of a baby's weight for six months

_____

2. to show how many bicycles were sold each month at a store

_____

3. to find the median age of the teachers at a school

_____

4. to compare the favorite sports of boys and girls in your class

_____

Explain why each graph or plot is not the best choice for the data it shows. Tell which type of graph or plot would be a better choice.

5.

Daily High Temperatures For Sept. 15–21

_____

_____

_____

6. **FAVORITE GAMES AT RECESS**

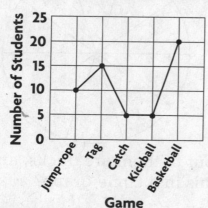

_____

_____

_____

_____

_____

Mixed Review

Complete to make the equation true.

7. $15 + 4 =$ _____ $+ 10$

8. _____ $+ 8 = 13 + 4$

9. $11 +$ _____ $= 20 + 15$

10. Find the value of $48 - (14 + 7)$. _____

Name _____

Problem Solving Skill
Draw Conclusions

For 1–7, use the graph.

The parents of Mrs. Watkins's fourth-grade students wanted to compare their favorite music choices for the Academic Dinner. Mrs. Watkins took a survey and made a double-bar graph.

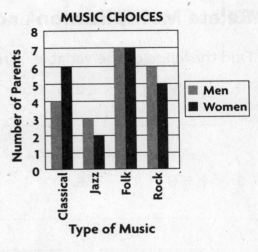

1. What is the favorite music choice for men?

   _____

2. What is the favorite music choice for women?

   _____

3. How many men prefer to have rock music at the banquet?

   _____

4. How many women prefer classical music?

   _____

5. Which type of music is preferred equally by the men and women?

   _____

6. How many men were surveyed altogether? women?

   _____

7. Is it reasonable to conclude that the parents chose folk music for the Academic Dinner? Explain.

   _____

   _____

Mixed Review

8. What number is 100,000 greater than 3,489,234?

   _____

9. Round 355,790 to the nearest thousand.

   _____

10. Estimate. 390,645 + 71,960

   _____

11. Estimate. 495,931 + 889,853

   _____

Relate Multiplication and Division

Find the value of the variable. Write a related equation.

1. $21 \div 3 = t$ **2.** $5 \times 5 = c$ **3.** $16 \div 2 = a$ **4.** $18 \div 6 = d$

_____   _____   _____   _____

_____   _____   _____   _____

5. $54 \div 9 = k$ **6.** $4 \times 4 = b$ **7.** $6 \times 2 = f$ **8.** $35 \div 7 = h$

_____   _____   _____   _____

_____   _____   _____   _____

9. $8 \div n = 2$ **10.** $4 \times p = 24$ **11.** $30 \div z = 6$ **12.** $6 \times j = 48$

_____   _____   _____   _____

_____   _____   _____   _____

13. $g \div 7 = 8$ **14.** $y \div 1 = 6$ **15.** $k \times 6 = 42$ **16.** $n \times 7 = 63$

_____   _____   _____   _____

_____   _____   _____   _____

Write the fact family for each set of numbers.

17. 3, 4, 12

_____

_____

18. 4, 7, 28

_____

_____

19. 5, 10, 50

_____

_____

20. 8, 9, 72

_____

_____

Mixed Review

21. $11.21
     $12.15
  +$ 1.61

22.   1,242,316
  −   164,320

23.   6,548,957
    3,847,200
  + 9,874,512

24. $15.27
    $ 7.99
  +$ 3.25

25.   8
  $\times 8$

26.   9
  $\times 4$

27.   6
  $\times 7$

28.   3
  $\times 5$

29.   7
  $\times 8$

Name _____

Multiply and Divide Facts Through 5

Write a related multiplication or division equation.

1. $2 \times 4 = 8$

_____

2. $2 \times 5 = 10$

_____

3. $2 \times 2 = 4$

_____

4. $4 \times 1 = 4$

_____

Find the product or quotient.

5. 2×6

6. $21 \div 7$

7. 5×9

8. $28 \div 4$

_____ _____ _____ _____

9. 3×8

10. $24 \div 6$

11. $18 \div 2$

12. 5×8

_____ _____ _____ _____

Find the value of the variable.

13. $2 \times 7 = 14$, so $(2 \times 7) + 10 = r$.

_____

14. $(36 \div 4) = 9$, so $(36 \div 4) \times 5 = m$.

_____

Write $<$, $>$, or $=$ in each $\bigcirc$.

15. $27 \div 3 \bigcirc 2 \times 4$

16. $32 \div 4 \bigcirc 3 \times 3$

Mixed Review

17. Find the value. $(22 - 6) + 38$

_____

18. In the number 1,257,863 what digit is in the ten thousands place?

_____

Multiply and Divide Facts Through 10

Show how the arrays can be used to find the product.

1. What is 7 × 8?

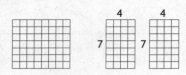

7 × 4 = _____

7 × 4 = _____

So, 7 × 8 = _____ .

2. What is 6 × 8?

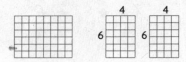

6 × 4 = _____

6 × 4 = _____

So, 6 × 8 = _____ .

Find the product or quotient. Show the strategy you used.

3. 6 × 6

_____

4. 56 ÷ 7

_____

5. 8 × 5

_____

6. 36 ÷ 4

_____

7. 10 × 6

_____

8. 72 ÷ 8

_____

9. 9 × 7

_____

10. 56 ÷ 8

_____

11. 8 × 6

_____

12. 42 ÷ 6

_____

13. 90 ÷ 9

_____

14. 9 × 9

_____

15. 7 × 6

_____

16. 8 × 9

_____

17. 49 ÷ 7

_____

18. 54 ÷ 9

_____

Mixed Review

19. In the number 125,588,325 what digit is in the ten millions place?

_____

20. Find the elapsed time.
Start: 7:54 A.M. End: 9:12 P.M.

_____

21. Round 362,847,321 to the nearest million.

_____

22. Round 13,567 to the nearest hundred.

_____

23. Write an expression using the variable *n*. There were 9 pears in the bowl. Jenny took some out.

_____

24. Write an equation using the variable *p*. Ed had some pens. He gave Ben 6 and now has 12.

_____

Multiplication Table Through 12

Vocabulary

Complete the sentence.

×	0	1	2	3	4	5	6	7	8	9	10	11	12
0	0	0	0	0	0	0	0	0	0	0	0	0	0
1	0	1	2	3	4	5	6	7	8	9	10	11	12
2	0	2	4	6	8	10	12	14	16	18	20	22	24
3	0	3	6	9	12	15	18	21	24	27	30	33	36
4	0	4	8	12	16	20	24	28	32	36	40	44	48
5	0	5	10	15	20	25	30	35	40	45	50	55	60
6	0	6	12	18	24	30	36	42	48	54	60	66	72
7	0	7	14	21	28	35	42	49	56	63	70	77	84
8	0	8	16	24	32	40	48	56	64	72	80	88	96
9	0	9	18	27	36	45	54	63	72	81	90	99	108
10	0	10	20	30	40	50	60	70	80	90	100	110	120
11	0	11	22	33	44	55	66	77	88	99	110	121	132
12	0	12	24	36	48	60	72	84	96	108	120	132	144

1. A _____ is the product of a given number and another whole number.

Use the multiplication table to find the product or quotient.

2. $40 \div 4$

3. 5×10

_____ _____

4. $70 \div 10$

5. $110 \div 10$

6. $11 \div 1$

7. 10×8

_____ _____ _____ _____

8. 7×12

9. $108 \div 9$

10. 11×5

11. $36 \div 3$

_____ _____ _____

Find the value of the variable.

12. $30 \div 10 = t$

13. $121 \div y = 11$

14. $80 \div 8 = h$

15. $n \times 12 = 48$

_____ _____ _____ _____

16. What are the multiples of 8 shown in the table?

_____

Mixed Review

17. In 7,894,132, what digit is in the ten thousands place?

_____

18. Round 639 to the nearest ten.

_____

19. Find the median.
15, 18, 22, 11, 20, 20, 13

_____

20. Find the mode.
15, 18, 22, 11, 20, 20, 13

_____

21. $(14 - 8) + 17 = $ _____

22. $36 - (3 + 9) = $ _____

23. $(15 + 15) - (12 + 2) = $ _____

24. $(17 - 6) + (42 - 17) = $ _____

Name _____

Multiplication Properties

Vocabulary

Word Bank
Commutative   Zero
Identity

Complete the sentence.

1. The _____ Property of Multiplication states
that the product of 1 and any number is that number.

2. The _____ Property of Multiplication states
that the product of 0 and any number is 0.

3. The _____ Property of Multiplication states that two numbers can be multiplied in any order and the answer will be the same.

Find the missing number. Name the property you used.

3. $2 \times (4 \times 3) = (2 \times 4) \times$ __   4. $12 \times$ __ $= 12$ 5. $10 \times$ __ $= 0$

_____   _____   _____

Show two ways to group by using parentheses. Find the product.

6. $2 \times 2 \times 5$ 7. $3 \times 4 \times 3$ 8. $2 \times 4 \times 5$

_____   _____   _____

Write the missing number. Then find the product.

9. $5 \times 3 = 3 \times$ _____   10. $4 \times 7 =$ _____ $\times 4$ 11. $9 \times$ _____ $= 8 \times 9$

_____   _____   _____

Write $<$, $>$, or $=$ in each $\bigcirc$.

12. $6 \times (4 \times 2) \bigcirc (2 \times 3) \times 8$ 13. $5 \times (4 \times 3) \bigcirc (3 \times 3) \times 6$

Mixed Review

14. Write one thousand, eighty-five in standard form. _____

15. $\$63 + \$48 + \$122$ 16. $\$104 - \$27$ 17. $\$83 + \$52 + \$36$

_____   _____   _____

Problem Solving Skill

Choose the Operation

Solve. Name the operation or operations you used.

1. Kate sold 21 boxes of cookies. Randy sold 32 boxes of cookies. Gina sold 49 boxes of cookies. How many boxes did they sell?

2. Behind home plate there are 5 rows of seats. Each row has 7 seats in it. How many seats are in this section?

3. The pottery classroom has 3 tables. There are 6 people at each table. If each person makes 2 clay animals, how many clay animals are made?

4. The fine for an overdue book at the Cotter Library is 5¢ a day. Tyler returned his books 1 day late. He paid a 30¢ fine. How many books did he return?

5. Ashley, Suzanne and Liz bought a box of chocolates. There are 36 chocolates in the box. How many do they get each?

6. Clyde sleeps 8 hours each night. How many hours does he sleep each week?

7. On Tuesday morning, Mrs. Corbett drove 57 miles to Princeton. Then she drove to Natick. She drove a total of 90 miles. How many miles was it from Princeton to Natick?

8. Peter took a three-day 28-mile backpacking trip. He hiked 9 miles the first day and 11 miles the second day. How many miles did he hike the third day?

Mixed Review

9. Find the median.
546, 550, 420, 410, 560, 530, 530

10. Find the mode.
546, 550, 420, 410, 560, 530, 530

11. In the number 12,482 what digit is in the tens place?

12. What is the elapsed time between 9:27 A.M. and 6:32 P.M.?

© Harcourt

Expressions

Find the value of the expression.

1. $(3 \times 8) + 5$ **2.** $4 + (25 \div 5)$ **3.** $(7 \times 3) - 4$ **4.** $8 + (9 \times 6)$

_____   _____   _____   _____

5. $(18 \div 6) - 3$ **6.** $34 - (4 \times 4)$ **7.** $(28 \div 7) + 3$ **8.** $4 \times (6 + 3)$

_____   _____   _____   _____

Write an expression to match the words.
Then find the value of the expression.

9. Tyler has $25. She wants to buy
4 movie tickets at $3 each for
herself and her friends. How
much money will she have
left over?

_____

10. Camille is decorating frames
with seashells. She makes
6 frames with 4 seashells each.
She has 3 shells left over. How
many shells did she start with?

_____

Mixed Review

11. 318
       $\underline{-\ 44}$

12. 502
       $\underline{-\ 27}$

13. $6,059$
       $\underline{-\ 479}$

14. $8,408$
       $\underline{-\ 4,629}$

15. Jordan wants to buy a new
video game. It will take him
5 months to save up enough
money. He begins saving
February 1. In what month will
he be able to buy the game?

_____

16. Megan needs to leave the
house at 7:30 A.M. She wants
to wake up to get ready 45
minutes before she leaves. What
time does she need to get up?

_____

17. Write 503,010 in words.

_____

_____

Name _____

Order of Operations

Vocabulary

Complete the sentence.

1. A special set of rules, called the _____,
   can be used to solve expressions with more than one
   operation.

Write *correct* if the order of operations are listed in the correct
order. If not, write the correct order of operations.

2. 72 ÷ (2 + 6)   Divide, add

3. 4 × (9 + 2)   Add, multiply

_____

_____

4. 42 − 18 ÷ 3   Divide, subtract

5. 12 − 6 × 4   Subtract, multiply

_____

_____

Follow the order of operations to find the
value of each expression.

6. 12 − 36 ÷ 9

7. 5 + 4 × 6

8. 8 × 6 + 4

_____

9. 45 ÷ 9 × 3

10. 2 × (42 ÷ 6)

11. (6 + 2) × 5

_____

12. 48 ÷ (4 × 2)

13. 18 ÷ 2 + 3 × 11

14. (8 + 12) ÷ 2 × 5

_____

Mixed Review

15.   409,558
        + 76,502

16.   23,141
        − 3,400

17.   530,120
        − 146,218

18.   805,844
        + 334,679

19.   8
        × 7

20.   7
        × 6

21.   9
        × 8

22.   4
        × 7

Expressions and Equations with Variables

Find the value of the expression.

1. $54 \div a$ if $a = 6$ **2.** $7 \times d$ if $d = 3$ **3.** $5 \times s$ if $s = 11$ **4.** $36 \div x$ if $x = 4$

_____ _____ _____ _____

Write an expression that matches the words.

5. 6 times the number of marbles, m, in a collection

_____

6. 7 pails of seashells times the number of shells, s, in each pail

_____

7. 72 children divided into equal groups, g

_____

8. 24 books divided equally between classrooms, c

_____

Write an equation for each. Choose a variable for the unknown.
Tell what the variable represents.

9. An equal number of people in each of 3 vans is a total of 18 people.

_____

_____

10. 36 bowlers divided equally among bowling lanes is 4 bowlers per lane.

_____

_____

Use mental math to solve each equation. Check your work.

11. $8 \times c = 32$ **12.** $a \times 9 = 45$ **13.** $42 \div s = 7$ **14.** $r \div 3 = 6$

_____ _____ _____ _____

Mixed Review

Round each number to the nearest thousand.

15. 7,605   **16.** 14,230   **17.** 39,824   **18.** 809,368

_____ _____ _____ _____

Problem Solving Strategy

Work Backward

Write an equation and *work backward* to solve.

1. Alexander had some nickels in his bank. He added 3 dimes to the bank and then he had 85¢. How many nickels did Alexander have?

_____

2. Roz is making a quilt. Yesterday she sewed some squares. Today she sewed together 3 rows with 10 squares each. She has sewn a total of 50 squares. How many squares did Roz sew yesterday?

_____

Work backward to solve.

3. Leo folded a sheet of paper in half a certain number of times. When unfolded, the sheet was divided into 8 sections. How many times did Leo fold the paper in half?

_____

4. Ann is setting a clock. It says 12:00 P.M. She moves the minute hand forward 10 minutes, back 12 minutes, forward 8 minutes, and back some minutes. If the time now reads 12:03 P.M., what was her final move?

_____

5. Holly is going from her home to the grocery store. To get to the store, she walks 3 blocks west and 2 blocks south. When she leaves the store, she walks 3 blocks east. How many blocks and in what direction should Holly walk to get home?

_____

6. Amy and Tim are playing a counting game. They are counting to 30. Amy claps when they say a number that can be divided evenly by 3. Tim claps when they say a number that can be divided evenly by 4. On what numbers do they both clap?

_____

Mixed Review

7.	**8.**	**9.**	**10.**	**11.**
3	9	9	12	12
$\times 8$	$\times 4$	$\times 9$	$\times 6$	$\times 10$

Balance Equations

Multiply or divide both sides by the given number. Find the new value.

1. 4 pennies = 4 pennies; multiply both sides by 7.

_____

2. 2 dimes = 2 dimes; multiply both sides by 3.

_____

3. 1 nickel = 5 pennies; multiply both sides by 7.

_____

4. 3 nickels = 1 dime 1 nickel; multiply both sides by 3.

_____

5. (4 + 2) = (3 × 2); divide both sides by 2.

_____

6. (6 + 3) = (3 × 3); divide both sides by 9.

_____

7. 12 = 6 × 2; divide both sides by 2.

_____

8. (3 + 5) = (64 ÷ 8); divide both sides by 4.

_____

9. 10 = 5 × 2; divide both sides by 5.

_____

10. (6 + 5) = (11 × 1); multiply both sides by 10.

_____

11. (2 + 3) = (15 ÷ 3); divide both sides by 5.

_____

12. 1 dime 2 pennies = 12 pennies; multiply both sides by 3.

_____

Mixed Review

Name the place value of the bold digit.

13. 1,**6**72,439 _____

14. 1,672,4**3**9 _____

15. 1,6**7**2,439 _____

16. **1**,672,439 _____

Solve.

17. $719.20
    +$48.44

18. 2,209
    −1,072

19. 4,476
    +4,467

20. $32.99
    − $12.81

Name _____

Patterns: Find a Rule

Find a rule. Write the rule as an equation. Find the missing number.

1.

Input	Output
a	b
15	■
20	4
25	5
30	6

2.

Input	Output
c	d
4	16
5	20
6	24
7	■

3.

Input	Output
s	t
■	16
3	24
4	32
5	40

4.

Input	Output
r	p
35	5
42	6
49	7
56	■

_____   _____   _____

_____   _____   _____

Use the rule and the equation to make an input/output table.

5. Multiply by 2. $a \times 2 = c$

Input	Output

6. Divide by 3. $r \div 3 = s$

Input	Output

7. Multiply by 11. $p \times 11 = q$

Input	Output

8. Divide by 4. $y \div 4 = z$

Input	Output

Mixed Review

Find the value of the expression.

9. 12×8

10. $99 \div 11$

11. $63 - (14 \div 7)$

_____   _____   _____

12. What time is 2 hours and 40 minutes after 11:22 A.M.?

13. Write the standard form for three hundred thousand, five.

_____   _____

Mental Math: Multiplication Patterns

Use mental math to complete.

1. $5 \times 5 =$ _____   2. $9 \times 8 =$ _____   3. $2 \times 3 =$ _____

   $5 \times 50 =$ _____ $9 \times 80 =$ _____ $2 \times 30 =$ _____

   $5 \times 500 =$ _____ $9 \times 800 =$ _____ $2 \times 300 =$ _____

   $5 \times 5,000 =$ _____ $9 \times 8,000 =$ _____ $2 \times 3,000 =$ _____

Use mental math. Write the basic fact and use a pattern to find the product.

4. 5×700 5. 9×400 6. 9×900

   _____   _____   _____

7. 4×500 8. $3 \times 4,000$ 9. $8 \times 3,000$

   _____   _____   _____

Find the value of n.

10. $6 \times 40,000 = n$ 11. $n = 3 \times 600$ 12. $n \times 500 = 3,500$

    _____   _____   _____

13. $3 \times n = 15,000$ 14. $n \times 8 = 640$ 15. $7 \times n = 42,000$

    _____   _____   _____

16. $7,000 \times n = 49,000$ 17. $6 \times n = 5,400$ 18. $n \times 6 = 1,800$

    _____   _____   _____

Mixed Review

19. Write the time in words.   20. Write the time in words.

_____   _____

_____   _____

Estimate Products

Vocabulary

Complete the sentence.

1. Numbers that are easy to compute with mentally are

   called _____.

Estimate the product. Choose the method.

2. 512
    × 5

3. 93
    × 8

4. 1,401
    × 7

5. 257
    × 3

6. 981
    × 7

7. 82
    × 4

8. 127
    × 9

9. 741
    × 9

10. $15.34 × 7

11. 903 × 4

12. 95 × 9

13. 718 × 3

_____   _____   _____   _____

14. 1,209 × 8

15. 657 × 3

16. 55 × 2

17. 9,099 × 4

_____   _____   _____   _____

Choose two factors from the box for each estimated product
□ × △.

309	4	759
193	3	7

18. 2,100 _____

19. 800 _____

20. 900 _____

21. 2,400 _____

22. 1,200 _____

23. 5,600 _____

Mixed Review

24. Order the numbers from *least* to *greatest*.

    182; 128; 1,028; 1,082

25. Round 194,012 to the nearest ten thousand.

_____   _____

Multiply 2-Digit Numbers

Find the product. Estimate to check.

1. $\begin{array}{r} 29 \\ \times\ 3 \\ \hline \end{array}$

2. $\begin{array}{r} 18 \\ \times\ 4 \\ \hline \end{array}$

3. $\begin{array}{r} 37 \\ \times\ 5 \\ \hline \end{array}$

4. $\begin{array}{r} 96 \\ \times\ 2 \\ \hline \end{array}$

5. $\begin{array}{r} 62 \\ \times\ 4 \\ \hline \end{array}$

6. $\begin{array}{r} 15 \\ \times\ 9 \\ \hline \end{array}$

7. $\begin{array}{r} 50 \\ \times\ 6 \\ \hline \end{array}$

8. $\begin{array}{r} 33 \\ \times\ 6 \\ \hline \end{array}$

9. 7×62

10. 4×86

11. 9×24

12. 3×78

Find the missing factor.

13. $2 \times \blacksquare = 52$

14. $3 \times \blacksquare = 135$

15. $7 \times \blacksquare = 203$

16. $8 \times \blacksquare = 392$

17. $\blacksquare \times 19 = 114$

18. $\blacksquare \times 99 = 297$

Compare. Write $<$, $>$, or $=$ in each $\bigcirc$.

19. $5 \times 15 \bigcirc 6 \times 12$

20. $3 \times 42 \bigcirc 6 \times 21$

21. $7 \times 22 \bigcirc 8 \times 17$

22. $9 \times 21 \bigcirc 6 \times 37$

23. $2 \times 79 \bigcirc 3 \times 24$

24. $8 \times 23 \bigcirc 4 \times 66$

Mixed Review

25. Which is greater, 909,872 or 990,678?

26. Round 192,875 to the nearest thousand.

27. Find the median.
    10, 12, 19, 18, 12, 13, 12

28. Find the mean.
    33, 36, 39, 45, 29, 58

Multiply 3- and 4-Digit Numbers

Find the product. Estimate to check.

1. 142×3

2. 263×4

3. 427×6

4. 538×7

5. $4{,}312 \times 2$

6. $1{,}726 \times 6$

7. $4{,}381 \times 8$

8. $2{,}473 \times 4$

9. 507×4

10. 352×2

11. 289×8

12. 345×6

13. $1{,}632 \times 5$

14. $2{,}658 \times 3$

15. $6{,}901 \times 2$

16. $3{,}146 \times 7$

Multiply. Then add to find the product.

17. 4×216

$(4 \times 200) + (4 \times 10) + (4 \times 6)$

_____

18. $2 \times 3{,}406$

$(2 \times 3{,}000) + (2 \times 400) + (2 \times 6)$

_____

Compare. Write $<$, $>$, or $=$ in each $\bigcirc$.

19. $6 \times 127 \bigcirc 2 \times 308$

20. $4 \times 755 \bigcirc 5 \times 604$

21. $3 \times 1{,}452 \bigcirc 2 \times 3{,}400$

Mixed Review

22. Write 93,702 in expanded form.

_____

23. If today is September 1, what was yesterday?

_____

24. The game started at 11:30 A.M. It ended at 12:15 P.M. How long did the game last?

_____

25. Tami and Tony each picked up a dozen eggs for their mother. How many eggs do they have in all?

_____

Multiply with Zeros

Find the product. Estimate to check.

1.   480
   $\times$ 3

2.   307
   $\times$ 4

3.   560
   $\times$ 6

4.   620
   $\times$ 7

5.   4,007
   $\times$ 2

6.   3,090
   $\times$ 6

7.   5,032
   $\times$ 8

8. $76.03
   $\times$ 4

9.   508
   $\times$ 4

10.   350
   $\times$ 7

11.   209
   $\times$ 8

12.   340
   $\times$ 6

13.   6,002
   $\times$ 5

14.   2,090
   $\times$ 3

15.   3,501
   $\times$ 8

16.   3,260
   $\times$ 7

Find the product. Estimate to check.

17. $(3 \times 507) \times 2$

18. $(4 \times \$10.09) \times 6$

19. $(7 \times 5,006) \times 3$

_____

_____

_____

20. $(5 \times \$23.90) \times 4$

21. $(2 \times 7,030) \times 5$

22. $(6 \times \$17.80) \times 4$

_____

_____

_____

Mixed Review

Find the value of n.

23. $n - 6 = 7$ _____

24. $3 + n = 8$ _____

25. $n \div 4 = 5$ _____

26. $n + 9 = 12$ _____

27. $9 + 11 = n$ _____

28. $n - 4 = 8$ _____

29. $n \times 4 = 28$ _____

30. $9 \times n = 36$ _____

31. $n + 7 = 12$ _____

32. $6 \times n = 48$ _____

33. $n - 2 = 6$ _____

34. $n \div 5 = 8$ _____

Problem Solving Strategy

Write an Equation

For 1–5, write an equation and solve.

1. Theresa's father works 5 days a week for 48 weeks a year. How many days does her father work in 1 year?

   _____

2. Theresa's father makes $24.50 per hour. How much does he make if he works 8 hours?

   _____

3. The football team is raising money for new footballs. How much money does the team need to raise if it wants 6 new footballs and each one costs $17.93?

   _____

4. A civil engineer counted the cars that passed through an intersection. If 2,457 cars passed through the intersection in one hour, how many cars would pass through the intersection in 8 hours?

   _____

5. Brianna practices playing guitar for 60 minutes each day. How many minutes does she practice in one week?

   _____

For 6–7, use this information.

Each floor of a nine-story office building has 132 windows.

6. What equation can you use to find the total number of windows?

   A $9 \times n = 132$ **C** $n \times 132 = 9$

   B $9 \times 132 = n$ **D** $n \times 9 = 132$

7. How many windows are there in all?

   F 188   **H** 1,088

   G 881   **J** 1,188

Mixed Review

8. $\begin{array}{r} 14 \\ \times\ 5 \\ \hline \end{array}$

9. $\begin{array}{r} 12 \\ \times\ 8 \\ \hline \end{array}$

10. $\begin{array}{r} 26 \\ \times\ 3 \\ \hline \end{array}$

11. $\begin{array}{r} 42 \\ \times\ 2 \\ \hline \end{array}$

12. $\begin{array}{r} 33 \\ \times\ 5 \\ \hline \end{array}$

13. $\$2.98 \times 7$ _____

14. $\$14.81 \times 3$ _____

Mental Math: Multiplication Patterns

Use a basic fact and a pattern to find the product.

1. $6 \times 5 =$ _____

   $6 \times 50 =$ _____

   $6 \times 500 =$ _____

2. $2 \times 2 =$ _____

   $2 \times 20 =$ _____

   $2 \times 200 =$ _____

3. $3 \times 6 =$ _____

   $3 \times 60 =$ _____

   $3 \times 600 =$ _____

   $3 \times 6{,}000 =$ _____

4. $9 \times 9 =$ _____

   $9 \times 90 =$ _____

   $9 \times 900 =$ _____

   $9 \times 9{,}000 =$ _____

5. $10 \times 3 =$ _____

   $10 \times 30 =$ _____

   $10 \times 300 =$ _____

   $10 \times 3{,}000 =$ _____

6. $40 \times 3 =$ _____

   $40 \times 30 =$ _____

   $40 \times 300 =$ _____

   $40 \times 3{,}000 =$ _____

7. $600 \times 30 =$ _____

8. $70 \times 3{,}000 =$ _____

9. $1{,}000 \times 30 =$ _____

10. $6{,}000 \times 6{,}000 =$ _____

Find the value of *n*.

11. $n \times 40 = 8{,}000$

12. $900 \times 300 = n$

_____

_____

Mixed Review

Round to the place value of the bold digit.

13. 57,**4**03,294

14. 9**8**3,204,448

15. 9**8**2,404

_____

_____

_____

Solve.

16.   300,010
   $-$ 255,492

17.   392,402
     392,402
  $+$ 492,148

18.   12,498
  $-$ 10,816

© Harcourt

The Distributive Property

1. Explain how to use the Distributive Property to find 20×12.

_____

_____

Use a model and the Distributive Property to find the product.

2. 6×19 **3.** 7×23 **4.** 9×21 **5.** 4×35

_____   _____   _____   _____

6. 10×16 **7.** 10×23 **8.** 10×19 **9.** 9×32

_____   _____   _____   _____

10. 20×14 **11.** 20×23 **12.** 50×11 **13.** 40×26

_____   _____   _____   _____

14. 50×12 **15.** 20×25 **16.** 12×10 **17.** 70×11

_____   _____   _____   _____

Mixed Review

Find the value of the expression.

18. $(7 + 2) - (5 + 1)$ **19.** $19 - (4 + 6)$ **20.** $5 + 8 + 6$

_____   _____   _____

21. $(3 + 3) \times 7$ **22.** $(12 - 3) \times 8$ **23.** $4 \times 2 \times 3$

_____   _____   _____

Multiply by Tens

Find the product.

1. 30
    × 5

2. 60
    × 30

3. 85
    × 30

4. 67
    × 90

5. 30
    × 70

6. 80
    × 5

7. 82
    × 50

8. 95
    × 50

9. 74 × 20

_____

10. 50 × 48

_____

11. 60 × 29

_____

12. 93 × 40

_____

13. 28 × 50

_____

14. 72 × 90

_____

Find the missing digits.

15. 30 × _____0 = 300

16. _____0 × 20 = 800

17. 16 × _____0 = 640

18. 4_____ × 80 = 3,600

19. 1_____ × 30 = 540

20. _____4 × 50 = 3,200

21. 8_____ × 20 = 1,700

22. 9_____ × 60 = 5,700

23. _____6 × 80 = 6,080

Mixed Review

Solve.

24. $n \times 4 = 28$

_____

25. $81 \div b = 9$

_____

26. $t \times (3 \times 2) = 18$

_____

27. $y \times 60 = 420$

_____

28. $300 \times w = 36,000$

_____

29. $p \times 500 = 6,000$

_____

30. 13
     × 4

31. 21
     × 5

32. 17
     × 2

33. 18
     × 5

34. 19
     × 3

35. 25
     × 4

36. 16
     × 8

37. 14
     × 7

Estimate Products

Round each factor. Estimate the product.

1. 35
    × 11

2. 54
    × 32

3. 97
    × 93

4. 549
    × 65

5. 486
    × 74

6. 658
    × 209

7. 648
    × 174

8. 840
    × 151

9. 339
    × 359

10. 884
    × 444

11. 312
    × 45

12. 951
    × 84

13. 503
    × 49

14. 320
    × 40

15. 503
    × 39

16. 85 × 81

17. 814 × 242

18. 957 × 84

19. 584 × 394

20. 84 × 315

Estimate to compare. Write <, >, or = in each ◯.

21. 609 × 43 ◯ 20,000

22. 15,000 ◯ 459 × 35

23. 872 × 254 ◯ 300,000

24. 965 × 19 ◯ 40,000

Mixed Review

Estimate by rounding to the greatest place value.

25.   485,492
    − 39,492

26.   493,430
       483,582
    + 7,302,598

27. 361
    × 42

28. 729
    × 58

Multiply.

29. 4,000
    ×   70

30. 900
    × 300

31. 6,000
    × 200

32. 3,200
    ×   20

Problem Solving Strategy

Solve a Simpler Problem

Break the problem into simpler parts and solve.

1. $40 \times 28 = (40 \times 20) + (40 \times 8)$

   $= \underline{\hspace{1.5cm}} + \underline{\hspace{1.5cm}}$

   $= \underline{\hspace{1.5cm}}$

2. $80 \times 49 = (\underline{\hspace{0.8cm}} \times \underline{\hspace{0.8cm}}) +$

   $(\underline{\hspace{0.8cm}} \times \underline{\hspace{0.8cm}})$

   $= \underline{\hspace{1.5cm}} + \underline{\hspace{1.5cm}}$

   $= \underline{\hspace{1.5cm}}$

A warehouse has many pieces of wood in stock. It is going to sell 312 bundles of wood with 20 pieces of wood in each bundle. How many pieces of wood will be sold?

3. Write an expression to help you solve the problem.

4. Find the total number of pieces of wood sold.

During a bad storm, Benny is using candles for light. He has 30 candles and each one burns for about 115 minutes. About how many minutes of light will the candles give Benny?

5. Write an expression to help you solve the problem.

6. Find the number of minutes of light in 30 candles.

Mixed Review

7. Mr. Rawlins has 57 fourth graders in his classes. He gives them a test with 30 questions on it. How many answers will he have to read to grade papers?

8. Antoin has $12.00. He buys 12 pens that cost 80¢ each. How much money does he have left?

9. $(13 + 2) \times n = 60$

10. $12 - (3 \times 3) = y$

11. $(42 - 22) + x = 31$

Multiply 2-Digit Numbers

Use regrouping or partial products to find the product. Estimate to check.

1. $\begin{array}{r} 62 \\ \times 35 \\ \hline \end{array}$
2. $\begin{array}{r} 55 \\ \times 29 \\ \hline \end{array}$
3. $\begin{array}{r} 73 \\ \times 44 \\ \hline \end{array}$
4. $\begin{array}{r} 48 \\ \times 27 \\ \hline \end{array}$

5. $\begin{array}{r} 81 \\ \times 17 \\ \hline \end{array}$
6. $\begin{array}{r} 67 \\ \times 23 \\ \hline \end{array}$
7. $\begin{array}{r} 26 \\ \times 18 \\ \hline \end{array}$
8. $\begin{array}{r} 32 \\ \times 24 \\ \hline \end{array}$

9. $\begin{array}{r} \$74 \\ \times 16 \\ \hline \end{array}$
10. $\begin{array}{r} 69 \\ \times 36 \\ \hline \end{array}$
11. $\begin{array}{r} \$39 \\ \times 35 \\ \hline \end{array}$
12. $\begin{array}{r} 76 \\ \times 11 \\ \hline \end{array}$

13. $14 \times 53 =$ _____

14. $\$26 \times 77 =$ _____

15. $\$26 \times 74 =$ _____

16. $21 \times 79 =$ _____

Write the missing product.

17. $30 \times 19 = 570$,

    so $30 \times 18 =$ ☐

18. $65 \times 15 = 975$,

    so $65 \times 16 =$ ☐

19. $40 \times 21 = 840$,

    so $40 \times 22 =$ ☐

20. $36 \times 18 = 648$,

    so $36 \times 17 =$ ☐

Mixed Review

21. $\begin{array}{r} 29 \\ \times 5 \\ \hline \end{array}$
22. $\begin{array}{r} 17 \\ \times 4 \\ \hline \end{array}$
23. $\begin{array}{r} 38 \\ \times 9 \\ \hline \end{array}$
24. $\begin{array}{r} 52 \\ \times 8 \\ \hline \end{array}$
25. $\begin{array}{r} 91 \\ \times 3 \\ \hline \end{array}$

26. $\begin{array}{r} 42 \\ \times 7 \\ \hline \end{array}$
27. $\begin{array}{r} 59 \\ \times 8 \\ \hline \end{array}$
28. $\begin{array}{r} 32 \\ \times 4 \\ \hline \end{array}$
29. $\begin{array}{r} 75 \\ \times 9 \\ \hline \end{array}$
30. $\begin{array}{r} 52 \\ \times 6 \\ \hline \end{array}$

31. $12 \times 4 =$ _____

32. $8 \times 8 =$ _____

Multiply 3-Digit Numbers

Find the product. Estimate to check.

| 1. | 221
 × 17 | 2. | $447
 × 36 | 3. | 727
 × 32 | 4. | 362
 × 27 |

| 5. | 549
 × 22 | 6. | $7.29
 × 46 | 7. | 636
 × 34 | 8. | 659
 × 73 |

9. 74 × 138 = _____

10. 25 × 808 = _____

11. 89 × $465 = _____

12. 19 × $517 = _____

Find the value for *n* that makes the equation true.

13. *n* × 720 = 10,800

14. 491 × *n* = 8,838

15. *n* × 679 = 5,432

_____   _____   _____

Mixed Review

16. (25 ÷ 5) + 10

17. 40 ÷ (2 × 4)

18. (48 ÷ 8) × (3 + 8)

_____   _____   _____

19. (36 ÷ 4) + (12 × 5)

20. (15 × 3) − (56 ÷ 8)

21. (19 + 44) ÷ 7

_____   _____   _____

| 22. | 6,442
 + 2,192 | 23. | 4,612
 − 895 | 24. | 3,292
 − 2,890 | 25. | 6,505
 − 398 |

| 26. | 70
 × 5 | 27. | 25
 × 6 | 28. | 35
 × 8 | 29. | 40
 × 5 | 30. | 15
 × 7 |

Name _____

Choose a Method

Find the product. Estimate to check.

1. 2,001
   ×   96

2. $2,425
   ×   24

3. 3,478
   ×   47

4. $5,699
   ×   26

5. 1,527
   ×   76

6. 3,639
   ×   69

7. 7,498
   ×   55

8. 6,643
   ×   78

9. 48 × 2,769 = _____

10. 36 × 4,873 = _____

Exercises 11–12 show 2 common errors. Describe each
error and correct it.

11. 1,360
    ×   42
    ─────
     272
    5,440
    ─────
    5,712

12. 2,966
    ×   16
    ──────
    17,796
    29,660
    ──────
    36,356

_____

_____

_____

_____

Mixed Review

13. (4 × 7) × 5

14. (6 × 10) × 2

15. (40 ÷ 8) × 12

_____

_____

_____

16. 19
    × 60

17. 29
    × 11

18. 32
    × 28

19. 2,511
    ×   16

20. 787
    − 319

21. 4,612
    −   895

22. 3,292
    − 2,890

23. 6,908
    − 5,002

Practice Multiplication Using Money

Find the product. Estimate to check.

1. $2,091
    × 26

2. $5.84
    × 6

3. $518
    × 27

4. $3.20
    × 84

5. $3,493
    × 36

6. $45.39
    × 31

7. $2,949
    × 26

8. $813
    × 63

9. $40.30
    × 64

10. $5,403
    × 38

11. $942
    × 81

12. $3,009
    × 49

Mixed Review

13. School ended at 3:20 P.M. Ida walked to Sam's house, which took 20 minutes. She stayed there for 1 hour. Then she had to walk home. The walk from Sam's house to her home took 40 minutes. At what time did she get home?

14. Marilu's dad has some weights in the basement. Marilu is trying to lift a box with three 5-lb weights, seven 1-lb weights, and two 7-lb weights. How much weight is in the box?

Complete the table.

15.

×	5	7	2	8	3	9	12	6
12								

16. 10,000
    − 5,794

17. 25,000
    − 21,211

18. 19,000
    − 9,655

19. 31,000
    − 28,414

Name _____

Problem Solving Skill

Multistep Problems

For 1–4, use the table.

The school cafeteria can add two new meals to the menu. They have been testing four meals and will choose the one that is most popular and the one that made the most money. The table shows the number of students who ate each meal and the price of the meal.

LUNCH MENU		
Food	Number of Students	Price of Each Meal
chicken patties	302	$1.12
veggie burger	309	$0.89
cheese sandwich	307	$0.95
tomato soup	189	$1.05

1. Write an expression to find the amount of money brought in by veggie burgers.

_____

2. How much money is brought in by sales of tomato soup?

_____

3. How much more money is brought in by chicken patties than by cheese sandwiches?

_____

4. Which two new meals will the cafeteria staff choose?

_____

_____

Mixed Review

5. $12.27
   ×   3

6. $8.99
   ×   4

7. $11.15
   − $7.27

8. $19.89
   − $6.40

9. $65 \times (437 - 81) = n$

10. $312 \times n = 24{,}336$

_____

_____

Divide with Remainders

Vocabulary

1. In a division problem, the _____ is the
amount left over when a number cannot be divided evenly.

_____

Make a model, record, and solve.

2. $4\overline{)19}$ 3. $3\overline{)25}$ 4. $6\overline{)38}$ 5. $2\overline{)17}$

Divide. You may wish to use counters.

6. $7\overline{)61}$ 7. $5\overline{)47}$ 8. $3\overline{)19}$ 9. $8\overline{)43}$

10. $6\overline{)58}$ 11. $9\overline{)49}$ 12. $2\overline{)13}$ 13. $7\overline{)65}$

Mixed Review

Complete each table.

14.
×	4	5	9	3	11	7	6
6							

15.
×	11	12	5	8	7	4	6
12							

Name _____

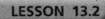

Model Division

Make or draw a model. Record and solve.

1. $52 \div 3 =$ _____   2. $68 \div 4 =$ _____   3. $65 \div 5 =$ _____

4. $7\overline{)91}$ 5. $6\overline{)100}$ 6. $2\overline{)58}$

7. $63 \div 3 =$ _____   8. $78 \div 4 =$ _____   9. $53 \div 4 =$ _____

10. $2\overline{)38}$ 11. $3\overline{)48}$ 12. $6\overline{)72}$

Mixed Review

For 13–15, use the table. The students in Mr. Jackson's class are holding a bake sale.

BAKE SALE ITEMS	
Kind of Cookie	Total Number
Chocolate chip	42
Oatmeal	65
Ginger	48

13. If Sara divides the chocolate chip cookies evenly into 3 bags, how many cookies does she put into each bag?

_____

14. If Tim divides the oatmeal cookies evenly into 5 bags, how many cookies does he put into each bag?

_____

15. Mr. Brown bought one bag of cookies for $1.75. What change should he receive from $10.00?

_____

Find the sum or difference.

16.　　$17.50
　　 + $17.50
　　 _____

17.　　$248.32
　　 − $119.55
　　 _____

18.　　$49.68
　　 − $5.11
　　 _____

19.　　$22.99
　　 + $85.98
　　 _____

Division Procedures

Divide and check.

1. 2)64   Check: 2. 3)96   Check: 3. 4)51   Check:

4. 3)94   Check: 5. 7)93   Check: 6. 8)89   Check:

Mixed Review

7. Shari sold 114 boxes of cookies with 14 cookies in each box. How many cookies did she sell?

_____

8. A football stadium can seat 50,013 people. If 24,394 seats are empty, how many people are attending the game?

_____

9. $8 \times 9 = 72$

$9 \times 8 =$ _____

$72 \div$ _____ $= 8$

_____ $\div 8 = 9$

10. $12 \times 7 =$ _____

$7 \times 12 =$ _____

$84 \div 7 =$ _____

$84 \div 12 =$ _____

11. $7 \times 6 =$ _____

_____ $\times 7 = 42$

$42 \div 7 =$ _____

_____ $\div 6 =$ _____

Problem Solving Strategy

Predict and Test

Predict and test to solve.

1. There were 93 students going to a nature camp. After equal groups of fewer than 10 students were formed for hiking, 2 students were left over. How many equal groups were formed?

   _____

2. During a hike, Sally and Dave collected 160 acorns. Sally collected 3 times as many acorns as Dave. How many acorns did Dave collect?

   _____

3. The 93 nature camp students ate lunch at the lodge. They sat at an even number of tables. There were 5 students sitting at one table, and an equal number of students sitting at each of the other tables. How many students were sitting at each of the other tables?

   _____

4. At one table, some of the students shared 3 pizzas. Each pizza was cut into 8 slices. After the students shared the pizza equally, there were 3 slices left over. How many students shared the pizza? How many slices of pizza did each student eat?

   _____

Mixed Review

For 5–8, use the graph.

5. For which candidate is the difference between the number of men's and women's votes the greatest?

   _____

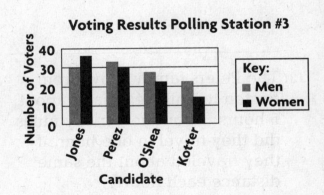
Voting Results Polling Station #3

6. About how many women voted for Jones?

   _____

7. About how many men voted for O'Shea?

   _____

8. About how many people voted at Polling Station #3? _____

Mental Math: Division Patterns

Use a basic division fact and patterns to write each quotient.

1. $240 \div 6 =$ _____

2. $350 \div 5 =$ _____

3. $360 \div 4 =$ _____

$2,400 \div 6 =$ _____

$3,500 \div 5 =$ _____

$3,600 \div 4 =$ _____

$24,000 \div 6 =$ _____

$35,000 \div 5 =$ _____

$36,000 \div 4 =$ _____

Divide mentally. Write the basic division fact and the quotient.

4. $210 \div 3$

5. $2,700 \div 3$

6. $8,000 \div 2$

7. $450 \div 9$

_____ _____ _____ _____

8. $40,000 \div 8$

9. $3,200 \div 8$

10. $120 \div 10$

11. $36,000 \div 6$

_____ _____ _____ _____

Mixed Review

For Problems 12–14, use the table.

12. The Shaw family drove from Boston to Houston in 6 days. If they drove about the same distance each day, about how many miles did they drive each day?

ROAD MILEAGE FROM BOSTON, MA	
City	Number of Miles
Kansas City, MO	1,391
Philadelphia, PA	296
Houston, TX	1,804

_____

13. The Peters family drove from Boston to Philadelphia in about 6 hours. About how many miles did they travel in one hour, if they traveled about the same distance each hour?

_____

14. Tom and his family leave Boston on Monday morning to drive to Kansas City. If they drive about 200 miles each day, what day should they arrive in Kansas City?

_____

Name _____

Estimate Quotients

Choose the letter of the best estimate.

1. $39 \div 5$ **a.** 70 or 80   **b.** 7 or 8   **c.** 15 or 20

2. $715 \div 7$ **a.** 17 or 18   **b.** 10 or 11   **c.** 100 or 110

3. $156 \div 4$ **a.** 12 or 13   **b.** 40 or 50   **c.** 4 or 5

Estimate using compatible numbers.

4. $2\overline{)17}$ 5. $4\overline{)23}$ 6. $6\overline{)375}$ 7. $8\overline{)255}$

8. $5\overline{)2,681}$ 9. $4\overline{)3,289}$ 10. $8\overline{)4,007}$ 11. $3\overline{)1,811}$

12. $3\overline{)241}$ 13. $5\overline{)4,787}$ 14. $5\overline{)388}$ 15. $7\overline{)3,594}$

Mixed Review

16. $2 \times 7 \times 2 =$ _____   17. $9 \times 5 \times 1 =$ _____   18. $2 \times 4 \times 7 =$ _____

19. $12 - 2 =$ _____ $+ 5$ 20. $9 \times 9 =$ _____ $\div 2$

21. $20 +$ _____ $= 16 + 24$ 22. $11 \times 6 =$ _____ $\div 3$

23. $\begin{array}{r} \$1,572 \\ - \$\ \ 803 \\ \hline \end{array}$ 24. $\begin{array}{r} 62,109 \\ - 45,863 \\ \hline \end{array}$ 25. $\begin{array}{r} \$1,438 \\ + \$5,760 \\ \hline \end{array}$ 26. $\begin{array}{r} 1,990 \\ + 3,473 \\ \hline \end{array}$

27. $\begin{array}{r} \$75,983 \\ + \$56,424 \\ \hline \end{array}$ 28. $\begin{array}{r} 371,812 \\ + 142,963 \\ \hline \end{array}$ 29. $\begin{array}{r} 400,000 \\ - 278,896 \\ \hline \end{array}$ 30. $\begin{array}{r} \$608,220 \\ - \$375,922 \\ \hline \end{array}$

Place the First Digit

Tell where to place the first digit. Then divide.

1. 5)36 _____

2. 3)62 _____

3. 2)173 _____

4. 6)72 _____

5. 4)241 _____

6. 7)702 _____

7. 9)381 _____

8. 4)820 _____

Divide.

9. 6)45

10. 3)84

11. 5)149

12. 7)652

13. 2)157

14. 3)171

15. 7)823

16. 8)799

Mixed Review

17.   32
     × 12

18.   48
     × 11

19.   288
     ×   5

20.   534
     ×   8

21.   4,211
    + 1,399

22.   2,378
    + 2,564

23.   5,913
    − 2,708

24.   25,926
    − 15,827

Divide 3-Digit Numbers

Divide.

1. $4\overline{)137}$

2. $3\overline{)325}$

3. $2\overline{)198}$

4. $7\overline{)924}$

Divide and check.

5. $3\overline{)152}$ Check:

6. $2\overline{)542}$ Check:

7. $5\overline{)627}$ Check:

8. $324 \div 6 =$ _____   Check:

9. $647 \div 9 =$ _____   Check:

Mixed Review

10.
$$
\begin{array}{r}
14 \\
\times\ 25 \\
\hline
\end{array}
$$

11.
$$
\begin{array}{r}
348 \\
\times\ 55 \\
\hline
\end{array}
$$

12.
$$
\begin{array}{r}
4,542 \\
\times\ \ \ 17 \\
\hline
\end{array}
$$

13.
$$
\begin{array}{r}
351 \\
\times\ 84 \\
\hline
\end{array}
$$

14.
$$
\begin{array}{r}
8,421 \\
\times\ \ \ 20 \\
\hline
\end{array}
$$

15.
$$
\begin{array}{r}
2,621 \\
+\ 5,892 \\
\hline
\end{array}
$$

16.
$$
\begin{array}{r}
7,457 \\
-\ 3,329 \\
\hline
\end{array}
$$

17.
$$
\begin{array}{r}
\$29.82 \\
+\ 49.70 \\
\hline
\end{array}
$$

18.
$$
\begin{array}{r}
4,608 \\
-\ 3,789 \\
\hline
\end{array}
$$

19.
$$
\begin{array}{r}
4,816 \\
+\ 5,184 \\
\hline
\end{array}
$$

Zeros in Division

Write the number of digits in each quotient.

1. $4\overline{)364}$ 2. $6\overline{)612}$ 3. $3\overline{)411}$ 4. $7\overline{)105}$

5. $5\overline{)545}$ 6. $8\overline{)432}$ 7. $7\overline{)905}$ 8. $2\overline{)123}$

Divide.

9. $3\overline{)312}$ 10. $4\overline{)429}$ 11. $6\overline{)526}$ 12. $4\overline{)436}$

13. $6\overline{)724}$ 14. $5\overline{)531}$ 15. $9\overline{)250}$ 16. $7\overline{)903}$

Mixed Review

17. $8 \times 6 =$ _____   18. $12 \times 2 =$ _____   19. $9 \times 8 =$ _____

20. $4 \times 4 =$ _____   21. $6 \times 5 =$ _____   22. $7 \times 7 =$ _____

23. $7 \times 3 =$ _____   24. $9 \times 6 =$ _____   25. $12 \times 3 =$ _____

26. $11 \times 6 =$ _____   27. $3 \times 8 =$ _____   28. $8 \times 8 =$ _____

29. $9 \times 7 =$ _____   30. $12 \times 10 =$ _____   31. $5 \times 9 =$ _____

Name _____

Choose a Method

Divide.

1. $4\overline{)740}$ 2. $5\overline{)630}$ 3. $6\overline{)828}$ 4. $7\overline{)756}$ 5. $3\overline{)840}$

6. $9\overline{)945}$ 7. $4\overline{)840}$ 8. $2\overline{)734}$ 9. $8\overline{)400}$ 10. $7\overline{)483}$

11. $6\overline{)1,680}$ 12. $4\overline{)5,316}$ 13. $5\overline{)6,030}$ 14. $8\overline{)3,208}$ 15. $5\overline{)6,600}$

Mixed Review

16. Evaluate:
$(25 - 9) + (12 \div 3)$

17. Find the median:
3, 6, 4, 6, 3, 4, 6, 7, 2

18. Find the elapsed time.
Start time: 8:03 A.M.
End time: 2:51 P.M.

_____   _____   _____

19. $36 \times 12 =$ _____   20. $88 \times 11 =$ _____   21. $54 \times 9 =$ _____

© Harcourt

Problem Solving Skill

Interpret the Remainder

Solve. Tell how you interpret the remainder.

1. The 158 fourth graders from the Glenwood School are going on a picnic. If there are 8 hot dogs in a package, how many packages are needed for each student to have 2 hot dogs?

   _____

   _____

2. Some of the students baked cookies for the picnic. Jeff baked 50 cookies. How many packages of 3 cookies each could he make?

   _____

   _____

3. The 158 students divide up into teams of 8 for a scavenger hunt. The students who are left over form a smaller team. How many teams are there?

   _____

   _____

4. Mrs. Jackson bought 7 dozen eggs for an egg-tossing contest. If each of 26 teams is given the same number of eggs, how many eggs are left over?

   _____

   _____

Mixed Review

For 5–7, use the price list.

5. Kito bought 4 pencils, 2 erasers, and a ruler. How much money did he spend?

   _____

SCHOOL STORE PRICE LIST	
Item	**Price**
Pencil	$0.10
Eraser	$0.15
Ruler	$0.50

6. On Monday, the store sold 20 pencils, 10 erasers, and 3 rulers. On Tuesday, the store sold 15 pencils, 13 erasers, and 3 rulers. On which day did the store take in more money?

   _____

7. On Friday, the store received a new supply of 72 pencils. Bill arranged the new pencils in groups of 5. How many groups could he make? How many pencils were left over?

   _____

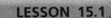

Name _____

Division Patterns to Estimate

Write the numbers you would use to estimate the quotient.
Then estimate.

1. 58 ÷ 15

_____

2. 695 ÷ 65

_____

3. 556 ÷ 68

_____

4. 273 ÷ 32

_____

5. 447 ÷ 52

_____

6. 810 ÷ 42

_____

Estimate.

7. 45 ÷ 14

_____

8. 362 ÷ 64

_____

9. 596 ÷ 34

_____

10. 79 ÷ 19

_____

11. 462 ÷ 83

_____

12. 721 ÷ 78

_____

Complete the tables.

	Dividend	Divisor	Quotient
13.	60	÷ 30	_____
14.	_____	÷ 30	20
15.	6,000	÷ 30	_____
16.	_____	÷ 30	2,000

	Dividend	Divisor	Quotient
17.	80	÷ 20	_____
18.	_____	÷ 20	40
19.	_____	÷ 20	400
20.	80,000	÷ 20	_____

Mixed Review

21.   39
       × 67

22.   379
       × 46

23.   3,593
       ×   4

24.   5,201
       ×   82

25. 81 ÷ 9 = _____

26. 140 ÷ 5 = _____

27. 320 ÷ 8 = _____

28. 72 ÷ 8 = _____

29. 660 ÷ 6 = _____

30. 490 ÷ 7 = _____

Model Division

Make a model to divide.

1. $15\overline{)67}$

2. $28\overline{)118}$

3. $21\overline{)85}$

4. $32\overline{)100}$

5. $35\overline{)176}$

6. $37\overline{)115}$

7. $78 \div 25 =$ _____

8. $97 \div 13 =$ _____

9. $117 \div 22 =$ _____

Use the model to complete the number sentence.

10.

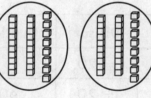

$61 \div 28 =$ _____

11.

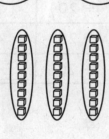

$38 \div 9 =$ _____

Mixed Review

12.
$$\begin{array}{r} 100{,}000 \\ \times\ \ \ \ 700 \\ \hline \end{array}$$

13.
$$\begin{array}{r} 495 \\ \times\ \ \ 39 \\ \hline \end{array}$$

14.
$$\begin{array}{r} \$872.64 \\ -\ \$41.98 \\ \hline \end{array}$$

15.
$$\begin{array}{r} \$784.32 \\ +\ \$32.53 \\ \hline \end{array}$$

16.
$$\begin{array}{r} 200{,}000 \\ \times\ 3{,}100 \\ \hline \end{array}$$

17.
$$\begin{array}{r} 702 \\ \times\ \ \ 44 \\ \hline \end{array}$$

18.
$$\begin{array}{r} \$90.89 \\ -\ \$89.77 \\ \hline \end{array}$$

19.
$$\begin{array}{r} \$645.30 \\ +\ \$822.98 \\ \hline \end{array}$$

Name _____

Division Procedures

Divide.

1. 22)598   **2.** 16)239   **3.** 11)346   **4.** 21)369

5. 13)461   **6.** 12)293   **7.** 31)862   **8.** 28)981

9. 17)206   **10.** 19)81   **11.** 23)485   **12.** 28)150

Mixed Review

13. 4)532   **14.** 4)626   **15.** 7)921   **16.** 4)5,881

17. 90,008 − 66,849	**18.** 967 × 56	**19.** 2,111 × 16	**20.** 72,931 + 30,275

© Harcourt

Correcting Quotients

Write *too high, too low,* or *just right* for each estimate. Then divide.

1. $17\overline{)152}^{\,8}$ _____

2. $35\overline{)186}^{\,4}$ _____

3. $42\overline{)351}^{\,7}$ _____

4. $48\overline{)374}^{\,8}$ _____

5. $52\overline{)419}^{\,7}$ _____

6. $76\overline{)679}^{\,8}$ _____

7. $63\overline{)556}^{\,9}$ _____

8. $67\overline{)650}^{\,9}$ _____

Mixed Review

9. Sue is packing 116 spools of thread into shoe boxes. Each box can hold 42 spools of thread. Will Sue be able to pack all the spools into 2 boxes? Explain.

10. Tony is estimating the time he needs to complete his math homework. He can complete about 3 problems per minute. If he allows 20 minutes, will he finish his 42 math problems? Explain.

Problem Solving Skill

Choose the Operation

Solve. Name the operation you used.

1. Mr. Murphy owns a bakery. On Saturday, he baked 60 blueberry muffins, 48 corn muffins, and 72 cranberry muffins. How many muffins did he bake in all?

_____

2. Mr. Murphy sold 498 cookies on Saturday. At the beginning of the day, there were 512 cookies. How many cookies were left at the end of the day?

_____

3. Susan bought 4 muffins for $0.79 each. How much money did she spend?

_____

4. Ryan paid $2.34 for 6 chocolate chip cookies. How much did each cookie cost?

_____

Mixed Review

For 5–7, use the graph.

5. How many bicycles were sold on Wednesday?

_____

6. How many bicycles were sold during the week?

_____

7. How many more bicycles were sold on Saturday than on Monday?

_____

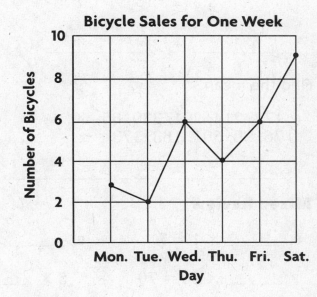

8. Will wants to buy a bicycle that costs $109. He has already saved $45. If Will earns $8 each week, how many weeks will it take him to save enough money to buy the bicycle?

9. Some days, Mary rides her bicycle to and from school. The distance is 2 miles each way. In October, Mary rode her bicycle to and from school 14 times. How many miles did she ride to and from school in October?

Find the Mean

Vocabulary

Complete.

1. A(n) _____ is the number found by dividing the sum of a set of numbers by the number of addends.

Write the division problem for finding the mean.
Then find the mean.

2.		3.			4.			5.	
7		3	16		143	352		2,516	
7		5	18		99	560		6,518	
10		6	19		213	184		3,215	
12		9	20		407	726		4,327	
14		12	4		698	938			
		13	7						

_____   _____   _____   _____

Find the mean.

6. 178, 214, 291, 339, 85, 59,   7. 9,972; 2,755; 1,130
   76, 99, 501, 163, 173

_____   _____

Mixed Review

8. _____ × 1 = 7   9. _____ × 4 = 20   10. 8 × _____ = 56

_____ × 10 = 70   5 × _____ = 200   _____ × 70 = 560

_____ × 100 = 700   5 × _____ = 2,000   8 × 700 = _____

11. 10 tens 5 ones = _____ tens 15 ones

12. 8 tens 17 ones = 9 tens _____ ones

13. 3 hundreds 14 tens = _____ hundreds 4 tens

14. 6 hundreds 2 tens = _____ hundreds 12 tens

Divisibility Rules

Vocabulary

Complete the sentence.

1. A number is _____ by another number when the quotient is a whole number and the remainder is zero.

___

Tell whether the number is divisible by 2, 3, 5, 9, or 10.

2. 36   **3.** 160   **4.** 225   **5.** 420

_____   _____   _____   _____

6. 189   **7.** 792   **8.** 1,080   **9.** 3,465

_____   _____   _____   _____

Write *true* or *false* for each statement. Explain.

10. All odd numbers are divisible by 3.

_____
_____
_____

11. All numbers that are divisible by 10 are divisible by 5.

_____
_____
_____

Mixed Review

12. $534 - 17$ **13.** $358 + 926$ **14.** $1,286 - 727$ **15.** $61,048 + 9,981$ **16.** $48,566 - 11,527$

17. $32,745 + 14,128$ **18.** $16,880 - 7,954$ **19.** $57,638 + 9,472$ **20.** $63,752 - 18,436$ **21.** $93,114 + 12,748$

Find the value of *n*.

22. $n \div 2 = 11$ _____   **23.** $6 \times n = 54$ _____   **24.** $28 \div n = 4$ _____

25. $132 \div n = 11$ _____   **26.** $n \times 7 = 56$ _____   **27.** $60 \div n = 12$ _____

28. $8 \times n = 6 \times 4$ _____   **29.** $n \times 4 = 6 \times 6$ _____   **30.** $n \div 5 = 30 \div 3$ _____

Factors and Multiples

Use arrays to find the factors of each number.

1. 12

2. 28

3. 45

_____   _____   _____

4. 32

5. 54

6. 33

_____   _____   _____

Find the first 6 multiples of each number.

7. 3 _____

8. 6 _____

9. 7 _____

10. 8 _____

11. 9 _____

12. 11 _____

Find the missing multiple.

13. 24, 32, ____, 48

14. ____, 42, 48, 54

15. 12, ____, 16, 18

Mixed Review

16. 16×7

17. 12×6

18. 18×6

19. 24×4

20. 11×11

21. $264 \div 2 =$ _____

22. $105 \div 5 =$ _____

23. $144 \div 9 =$ _____

24. $\$2.16 \times 3$

25. $\$8.45 \times 2$

26. $\$3.50 \times 3$

27. $\$2.25 \times 4$

28. $\$1.12 \times 6$

Prime and Composite Numbers

Make arrays to find the factors. Write *prime* or *composite* for each number.

1. 19 _____ 2. 32 _____ 3. 81 _____ 4. 36 _____

_____ _____ _____ _____

5. 27 _____ 6. 56 _____ 7. 29 _____ 8. 18 _____

_____ _____ _____ _____

Write *prime* or *composite* for each number.

9. 42 10. 64 11. 100 12. 72

13. 22 14. 15 15. 91 16. 47

Frances has to stack cans on a shelf. Each stack must have an equal number of cans. How many ways can she stack the cans on the shelf? List the ways.

17. | 18. | 19.

┌──────────────┐ ┌──────────────┐ ┌──────────────┐
│ **12 CANS** │ │ **24 CANS** │ │ **18 CANS** │
└──────────────┘ └──────────────┘ └──────────────┘

_____ _____ _____

_____ _____ _____

_____ _____ _____

_____ _____ _____

Mixed Review

20. Train A traveled the 29 miles between Dell City and Mesabi 18 times. Train B traveled the 21 miles between Mesabi and Dodge 24 times. Which train traveled the greater number of miles?

21. Joanna left school at 3:30 P.M. She went to volleyball practice for 90 minutes. She stopped at her aunt's house for 75 minutes, and then spent 15 minutes walking home. What time did she get home?

_____ _____

Name _____

Problem Solving Strategy

Find a Pattern

1. Continue the pattern.

1, 4, 7, 10, ___

_____

2. Continue the pattern.

3, 9, 27, 81, ___

_____

3. Describe the pattern in Exercise 1.

_____

4. Describe the pattern in Exercise 2.

_____

5. What could be the next two numbers in the following pattern?

1, 3, 7, 13, 21, ___, ___

_____

6. What could be the next two shapes in the following pattern?

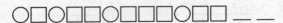 ___ ___

_____

7. Monica is playing a guessing game with her friends. When they say 5, she says 20. When they say 9, she says 36. When they say 2, she says 8. What is the pattern?

_____

8. Ruthie is writing a pattern where she gets a number by multiplying the last number by 2 and adding 3. Write the next two numbers.

1, 5, 13, 29, ___, ___

_____

Mixed Review

9. Melanie's family took a trip. The first day they drove 140 miles. The second day they drove 210 miles. The third day they drove 120 miles. The last day they drove 190 miles. What was their average mileage per day?

_____

10. Melanie's mother bought 30 gallons of gasoline during their trip. If they drove a total of 660 miles, how many miles did they drive per gallon of gasoline?

_____

11. If gasoline cost $1.45 per gallon, how much did Melanie's mother spend on gasoline for their trip?

_____

12. How much less would the total cost for gasoline have been if it had cost $1.25 per gallon?

_____

Square Numbers

Vocabulary

Complete the sentences.

1. A _____ is a number that is the product of any number and itself.

2. The _____ of a number is one of the two equal factors of that number.

Name the square number and the square root for each array.

3.

4.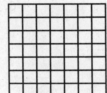

5.

6.

_____   _____   _____   _____

Find the square number.

7. 9×9 8. 2×2 9. 6×6 10. 8×8 11. 7×7

_____   _____   _____   _____   _____

Use a multiplication table to find the square root.

12. 100   13. 25   14. 1   15. 16   16. 144

_____   _____   _____   _____   _____

Mixed Review

17. $3 \times 4 \times 2 =$ _____   18. $5 \times 2 \times 4 =$ _____   19. $6 \times 2 \times 8 =$ _____

Name the place value of the bold digit.

20. 2,457,212 _____

21. 2,457,212 _____

22. 2,457,212 _____

23. 2,457,212 _____

Lines and Rays

Vocabulary

Fill in the blanks.

1. A _____ names an exact location in space.

2. A _____ is a straight path of points that continues without end in both directions. It has no endpoints.

3. A _____ is part of a line. It has one endpoint and continues without end in one direction.

Name a geometric term that describes each.

4. tip of a pen   5. a ruler   6. a playground

_____   _____   _____

Draw and label an example of each.

7. point *D*   8. line *MN*   9. ray *DE*

10. line segment *OP*   11. plane *CDE*   12. line *ST*

Draw each line segment with the given length.

13. $\overline{AD}$, 3 in.   14. $\overline{MN}$, 4 cm   15. $\overline{XY}$, $1\frac{1}{2}$ in.

Mixed Review

16. 9
    $\times 7$

17. 8
    $\times 6$

18. 6
    $\times 7$

19. 8
    $\times 8$

20. 10
    $\times 7$

21. $14 \div 2 =$ _____   22. $36 \div 6 =$ _____   23. $42 \div 6 =$ _____

© Harcourt

Measure and Classify Angles

Vocabulary

Fill in the blanks.

1. A _____ angle measures 90°.

2. An _____ angle measures greater than 0° and less than 90°.

3. An _____ angle measures greater than 90° and less than 180°.

Use a protractor to measure each angle.
Then write *acute, obtuse, right,* or *straight.*

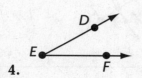

4. _____

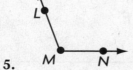

5. _____

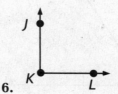

6. _____

Draw and label an example of each.

7. straight angle *QRS*   8. acute angle *XYZ*   9. obtuse angle *ABC*

Mixed Review

10.	11.	12.	13.	14.
9,824 + 3,201	8,729 − 5,597	693 + 478	3,091 − 1,482	12,643 − 7,365

15. 12 × 6 = _____   16. 11 × 8 = _____   17. 12 × 9 = _____

18. 50 ÷ 10 = _____   19. 60 ÷ 12 = _____   20. 44 ÷ 4 = _____

Line Relationships

Vocabulary

Fill in the blanks.

1. _____ lines cross each other at exactly one point and form four angles.

2. _____ lines intersect to form four right angles.

Name any line relationship you see in each figure. Write *intersecting*, *parallel*, or *perpendicular*.

3.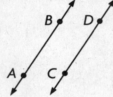

_____

4.

_____

5.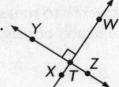

_____

6.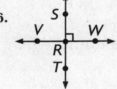

_____

7.

_____

8.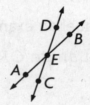

_____

Mixed Review

9. $4\overline{)22}$

10. $7\overline{)50}$

11. $9\overline{)14}$

12. $2\overline{)75}$

13. $\begin{array}{r} 17 \\ \times\ 15 \\ \hline \end{array}$

14. $\begin{array}{r} 259 \\ \times\ \ 5 \\ \hline \end{array}$

15. $\begin{array}{r} 78 \\ \times\ 9 \\ \hline \end{array}$

16. $\begin{array}{r} 361 \\ \times\ 20 \\ \hline \end{array}$

Problem Solving Strategy

Draw a Diagram

Draw a diagram on the grid to solve the problem.

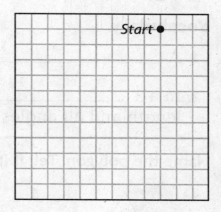

1. Trevor walks 10 units south on the Northern Trail. Then he walks 3 units west, 4 units north, 3 units west, and 6 units north. How much of the trail did Trevor walk?

   _____

2. Jason lives 5 blocks south of Kevin. Marcus lives 3 blocks north of Kevin. Tyler lives halfway between Jason and Marcus. Does Tyler live north or south of Kevin? How many blocks?

   _____

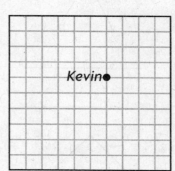

3. The city will construct a new building. From a starting point, its walls will go 8 units east, 5 units south, 4 units west, 2 units north, 4 units west, and 3 units north. The architect follows the path. How many units does he walk?

   _____

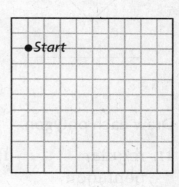

Mixed Review

Name any line relationship you see in each figure. Write *intersecting,* *parallel,* or *perpendicular.*

4.

5.

6.

_____   _____   _____

Draw and label an example of each.

7. point *T*   8. line *CD*   9. ray *FG*

Name _____

Polygons

Vocabulary

1. A _____ is a closed plane figure with straight sides.

2. In a _____ polygon, all sides have equal length and all angles have equal measure.

Name the polygon. Tell if it appears to be *regular* or *not regular*.

3.

_____

_____

4.

_____

_____

5.

_____

_____

6.

_____

_____

7.

_____

_____

8.

_____

_____

9.

_____

_____

10.

_____

_____

Draw each polygon.

11. regular pentagon

12. a triangle that is not regular

13. an octagon that is not regular

14. regular quadrilateral

Mixed Review

15. Name the line relationship.

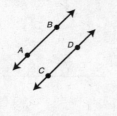

_____

16. Draw an example of line *AB*.

Name _____

Classify Triangles

Classify each triangle. Write *isosceles*, *scalene*, or *equilateral*.
Then write *right*, *acute*, or *obtuse*.

1.
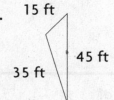
4 in.
6 in.
4 in.

_____

2.
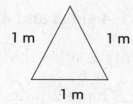
1 m
1 m
1 m

_____

3.
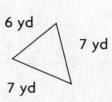
6 yd
7 yd
7 yd

_____

4.
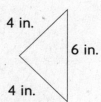
15 ft
45 ft
35 ft

_____

5.
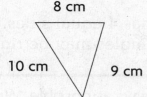
8 cm
10 cm
9 cm

_____

6.
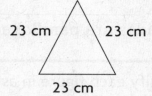
23 cm   23 cm
23 cm

_____

7.
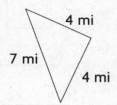
4 mi
7 mi
4 mi

_____

8.
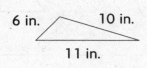
6 in.   10 in.
11 in.

_____

9.
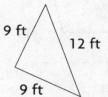
9 ft
12 ft
9 ft

_____

Classify each triangle by the length of its sides. Write *isosceles,
scalene,* or *equilateral.*

10. 12 in., 12 in., 12 in.   **11.** 65 yd, 43 yd, 65 yd   **12.** 45 mi, 23 mi, 56 mi

_____   _____

Mixed Review

Tell if the polygon is a *quadrilateral* or *not a quadrilateral.*

13.

_____

14.

_____

15.

_____

Name _____

Classify Quadrilaterals

Vocabulary

Fill in the blanks.

1. All _____ have 4 sides and 4 angles.

2. _____ have only 2 sides that are parallel.

3. _____ have 2 pairs of parallel sides. They may have 2 acute angles of the same size and 2 obtuse angles of the same size.

4. A _____ has 4 equal sides. Its opposite sides are parallel and its angles may be right angles.

Classify each figure in as many ways as possible. Write *quadrilateral, parallelogram, square, rectangle, rhombus,* or *trapezoid.*

5.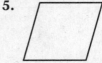

6.

7.

8.

_____   _____   _____   _____

_____   _____   _____   _____

Draw an example of each quadrilateral.

9. trapezoid   10. square   11. rhombus

12. parallelogram   13. rectangle   14. quadrilateral

Mixed Review

15. $\begin{array}{r} 250 \\ \times\ 7 \\ \hline \end{array}$

16. $\begin{array}{r} 864 \\ \times\ 5 \\ \hline \end{array}$

17. $\begin{array}{r} 793 \\ \times\ 6 \\ \hline \end{array}$

18. $\begin{array}{r} 122 \\ \times\ 8 \\ \hline \end{array}$

Problem Solving Strategy

Draw a Diagram

Use *draw a diagram* to solve.

1. Sort these figures into a Venn diagram showing *Figures with 4 Sides* and *Figures with More Than 4 Sides*: square, rectangle, pentagon, trapezoid, octagon, hexagon.

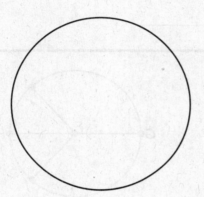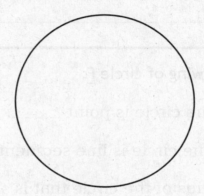

2. Sort these numbers into a Venn diagram showing *Divisible by 3* and *Divisible by 5*: 3, 5, 6, 10, 12, 15, 20, 24, 25, 30.

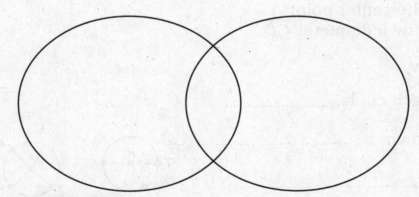

Mixed Review

Add or subtract.

3.	6,783	4.	8,743	5.	54,732	6.	9,275	7.	14,821
	+ 3,960		− 586		+ 4,694		+ 2,392		− 4,812

Circles

Vocabulary

Define the following words.

1. radius:_____

_____

2. diameter:_____

_____

For 3–6, use the drawing of circle <u>F</u>.

3. The center of the circle is point _____.

4. A diameter of the circle is line segment _____.

5. Name each radius of the circle that is shown.

   _____, _____, _____, and _____

6. Some points on the circle are _____, _____, _____, and _____.

7. Draw a circle. <u>L</u>abel the center point <u>A</u>.
   Draw a radius $\overline{AB}$. Draw a diameter $\overline{CD}$.

For 8–9, use circles *R* and *W*.

8. Name the center of each circle. _____

9. Name each radius shown. _____

   _____

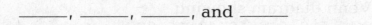

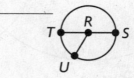

Mixed Review

10. A performance began at
    7:15 P.M. At 9:10 P.M., the
    performance ended. How long
    was the performance?

    _____

11. Rashid's bank has 6 quarters,
    9 dimes, 15 nickels, and
    26 pennies in it. How much
    is in his bank?

    _____

© Harcourt

Name _____

Turns and Symmetry

Tell whether the rays on the circle show a $\frac{1}{4}$, $\frac{1}{2}$, $\frac{3}{4}$, or complete turn.

1.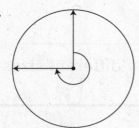

2.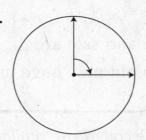

3.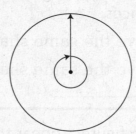

4.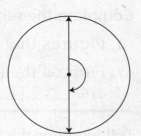

_____ _____ _____ _____

Tell whether the figure has *line symmetry, rotational symmetry,* or *both.*

5.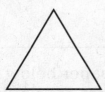

6.

7.

_____ _____ _____

Tell whether the figure has been turned 90°, 180°, 270°, or 360°.

8.

9.

10.

11.

_____ _____ _____ _____

Mixed Review

12. $3.78
    × 9

13. $10.50
    × 9

14. $689
    × 15

15. $187
    × 13

16. $345
    × 15

Divide.

17. 19)86

18. 34)139

19. 25)406

Congruent and Similar Figures

Vocabulary

Complete the sentences.

1. Figures that have the same shape and size are _____.

2. Figures that have the same shape but may have many different sizes are _____.

Tell whether the two figures appear to be *congruent*, *similar*, *both*, or *neither*.

3.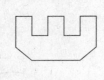

_____

4.

_____

5.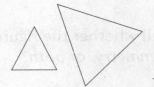

_____

6. Use the dot paper below. Draw two figures that are congruent.

7. Use the dot paper below. Draw two figures that are similar.

Mixed Review

8. $4,729 - 2,418 =$ _____

9. $2,470 - 981 =$ _____

10. $1,897 + 423 =$ _____

11. $6,231 + 4,865 =$ _____

12. $10,078 - 9,021 =$ _____

13. $9,624 - 3,071 =$ _____

14.
```
   738
   389
   388
 + 296
```

15.
```
   199
   309
   374
 + 902
```

16.
```
   422
   688
   201
 + 114
```

17.
```
   237
   640
   888
 + 315
```

© Harcourt

Problem Solving Strategy

Make a Model

For 1–4, make a model to solve.

1. Laura wants to make the figure below larger and then put it on her folder. Use 1-inch grid paper to help Laura make a larger picture.

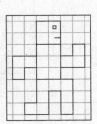

2. Wesley is decorating a bulletin board in his school hallway. He wants to make a larger picture of the figure below. Use 1-inch grid paper to help Wesley make the picture larger.

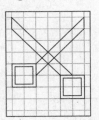

3. Make a smaller picture of the figure below. Use 0.5-cm grid paper to help you make a smaller picture.

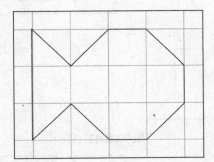

4. Make a larger picture of the figure below. Use 1-inch grid paper to help you.

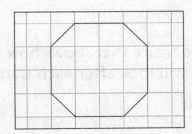

Mixed Review

5.
```
   589
+ 782
```

6.
```
  5,468
+ 9,230
```

7.
```
  10,860
－ 8,701
```

8.
```
  1,792
+ 4,567
```

9.
```
   907
－ 488
```

10.
```
   800
+ 745
```

11.
```
   3,459
－ 2,899
```

12.
```
  6,378
+ 8,719
```

13.
```
  6,448
－ 1,714
```

Transformations

Vocabulary

Complete the sentence.

1. The movement of a figure is a _____.

Tell how each figure was moved. Write *translation*, *reflection*, or *rotation*.

2.   **3.**    **4.**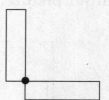

_____   _____   _____

5. Copy this figure on dot paper.
Then draw figures to show a translation,
a reflection, and a rotation.

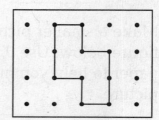

Circle the figure that shows how the figure at
the right will look after each transformation.

6. reflection   **7.** translation   **8.** rotation

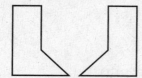   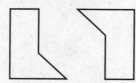   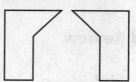

Mixed Review

Solve.

9. $(7 \times 6) \div 2 =$ ____   **10.** $(13 - 8) \times 9 =$ ____   **11.** $6 + (12 \div 2) =$ ____

12.   7,614
    + 8,093

13.   21,355
    − 9,787

14.   3,630
    × 41

15.   2,498
    × 15

Tessellations

Vocabulary

Complete the sentence.

1. A _____ is a pattern of closed figures that covers a surface with no gaps or overlaps.

Trace and cut out several of each figure. Tell if the figure will tessellate. Write *yes* or *no*.

2.
_____

3.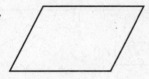
_____

4.
_____

Will these figures tessellate? Write *yes* or *no*.

5.
_____

6.
_____

7.
_____

Mixed Review

Write each number in expanded form.

8. 5,654 = _____ + _____ + _____ + _____

9. 9,232 = _____ + _____ + _____ + _____

10. 138,045 = _____ + _____ + _____ + _____ + _____

11. 87,657 = _____ + _____ + _____ + _____ + _____

Solve.

12. $(9 \times 4) \div 6 =$ _____

13. $20 - (12 \div 2) =$ _____

14. $8 + (5 \times 8) =$ _____

Geometric Patterns

Write a rule for the pattern. Then draw the next two figures.

1.

_____

2.

_____

3.

_____

Write a rule for the pattern. Then draw the missing figures.

4.   _____

_____

5.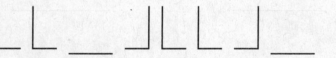

_____

Mixed Review

Solve.

6. 235	7. 656	8. 196	9. 378	10. 436
× 3	× 4	× 7	× 6	× 5

11. 21,582	12. 4,300	13. $83.72	14. 9,502	15. $39.17
+ 89,069	− 2,839	× 3	− 863	+ $68.39

© Harcourt

Temperature: Fahrenheit

Use the thermometer to find the temperature, in °F.

1.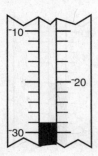

2.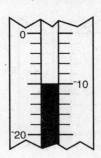

3.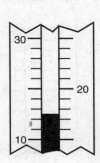

_____   _____   _____

For 4–7, use a thermometer to find the change in temperature.

4. 0°F to 35°F

5. ⁻10°F to 10°F

_____   _____

6. ⁻5°F to 25°F

7. ⁻15°F to 30°F

_____   _____

Circle the temperature that is a better estimate.

8. A pot of boiling water

10°F   or   212°F

9. A summer day in Florida

30°F or 95°F

10. An air-conditioned office building

75°F or 150°F

Mixed Review

Find the value of *n*.

11. $n \div 30 = 20$

12. $(25 + 5) - (10 \div 2) = n$

_____   _____

13. $n \times 6 = 72$

14. $88 \div n = 8$

_____   _____

15.   374
         + 129

16.   728
         + 152

17.   274
         − 186

18.   299
         − 119

Temperature: Celsius

Use the thermometer to find the temperature, in °C.

1.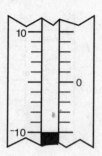

2.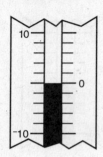

3.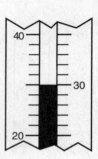

_____   _____   _____

For 4–7, use a thermometer to find the change in temperature.

4. 67°C to ⁻55°C

5. 48°C to ⁻10°C

_____   _____

6. ⁻1°C to 50°C

7. ⁻15°C to 22°C

_____   _____

Circle the temperature that is a better estimate.

8. the ice at the
ice rink

⁻1°C or 65°C

9. hot water in the
tea kettle

30°C or 100°C

10. a nice day for
a picnic

15°C or 80°C

Mixed Review

11. What is the change in
temperature, in °F, from the
boiling point (212°F) to the
freezing point (32°F) of water?

12. How are these odd numbers
alike? 5, 11, 17, 19, 23

_____   _____

13. 25)17,650   14. 22)12,056   15. 17)4,952   16. 29)511,607

Explore Negative Numbers

Use the number line to name the number each letter represents.

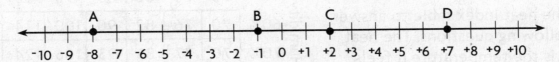

1. A = _____   **2.** B = _____   **3.** C = _____   **4.** D = _____

Draw a number line and graph each number and its opposite.

5. ⁻3

6. ⁺4

7. ⁺6

8. ⁻5

9. ⁻2

10. ⁺1

Compare. Write <, >, or = in each ◯.

11. ⁻8 ◯ ⁺2   **12.** ⁺8 ◯ ⁺2   **13.** 0 ◯ ⁺2   **14.** 2 ◯ ⁺2

15. ⁺9 ◯ ⁺2   **16.** ⁺1 ◯ ⁺8   **17.** 0 ◯ ⁻1   **18.** ⁻2 ◯ ⁺10

Mixed Review

19. List the factors of 18.

20. 36 × 100

21. What is the change in temperature from ⁻8°F to 8°F?

22. Which of these are composite numbers? 25, 31, 54, 79

Problem Solving Skill

Make Generalizations

Heat Index Table

Use the heat index table to answer the following questions. The heat index is the temperature it feels like, not the actual temperature.

Actual Temperature (°F)

Relative Humidity	70	75	80	85	90	95	100
60%	70	76	82	90	100	114	132
70%	70	77	85	93	106	124	144
80%	71	78	86	97	113	136	157
90%	71	79	88	102	122	150	170

1. Find the heat index for an outside temperature of 90°F with relative humidity of 70%.

   _____

2. What would the relative humidity be if it is 85°F with a heat index of 97°F?

   _____

3. What would the outside temperature be when the relative humidity is 90% with a heat index of 88°F?

   _____

4. Joe wants to take a walk. The relative humidity is 60% and the outside temperature is 75°F. Will it feel warmer or cooler than the outside temperature? Explain.

   _____

5. What generalizations can you make about the temperature that is read on the thermometer and the heat index?

   _____

   _____

Mixed Review

Compare. Write $<$, $>$, or $=$ in each ◯.

6. 7×9 ◯ $126 \div 2$

7. $^-7$ ◯ $^+5$

8. $^+9$ ◯ $^-8$

Subtract.

9. $\begin{array}{r} 56,703 \\ -\ 9,846 \\ \hline \end{array}$

10. $\begin{array}{r} 187,312 \\ -\ 74,961 \\ \hline \end{array}$

11. $\begin{array}{r} 836,031 \\ -248,712 \\ \hline \end{array}$

Name _____

Explore Inequalities
Vocabulary

Fill in the blank.

1. An _____ shows a relationship
   between two quantities that are not equivalent.

Which of the numbers 2, 3, 6, and 10 make each inequality true?

2. _____ < 9   3. _____ > 4   4. _____ > ⁻1

5. _____ ≤ 10   6. _____ ≥ 3   7. _____ + 2 > ⁺6

8. 2 + _____ ≤ 8   9. _____ + 5 ≥ 9   10. _____ − 2 ≤ 5

Draw a number line and graph three whole numbers that make
each inequality true.

11. _____ ≥ 3   12. _____ < 5

13. _____ + 4 > 6   14. 3 ≤ _____ − 1

15. 6 < _____ + 5   16. _____ + 2 ≥ 8

Mixed Review

17. 9 × 78   18. 4 × 386   19. 25 × 23   20. 6 × 2,308

_____   _____   _____   _____

21. 1,800 ÷ 3   22. 240 ÷ 6   23. 80 ÷ 4   24. 98 ÷ 5

_____   _____   _____   _____

25. $10.50   26. $24.53   27. 100,200   28. 46,623
   + 5.98   − 8.99   − 67,525   + 28,407

Name _____

Use a Coordinate Grid

Graph each ordered pair on the coordinate grid.

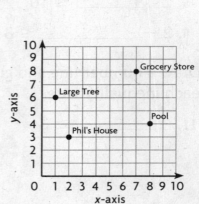

1. (1, 1)   2. (5, 4)

3. (8, 3)   4. (9, 9)

5. (8, 7)   6. (4, 6)

7. (3, 5)   8. (2, 7)

Write the ordered pair for each object on
the map.

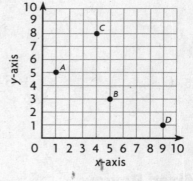

9. pool   10. Phil's house

_____   _____

11. grocery store   12. large tree

_____   _____

Write the ordered pair for each point on the
coordinate grid.

13. point *A*   14. point *B*

_____   _____

15. point *C*   16. point *D*

_____   _____

Mixed Review
Round each factor. Estimate the product.

17. 24 × 81 = _____   18. 36 × 52 = _____

19. 88 × 11 = _____   20. 45 × 219 = _____

21. 19 × 283 = _____   22. 72 × 72 = _____

23. 39 × 158 = _____   24. 18 × 18 = _____

Name _____

Read and Write Fractions

Vocabulary

Fill in the blank.

1. A number that names a part of a whole is a _____.

Write a fraction for the shaded part. Write a fraction for the unshaded part.

2.

3.

4.

5.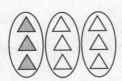

_____

_____

_____

_____

6.

7.

8.

_____

_____

_____

Draw a picture and shade part of it to show the fraction. Write a fraction for the unshaded part.

9. $\frac{2}{6}$

10. $\frac{7}{8}$

11. $\frac{4}{5}$

_____

_____

_____

Mixed Review

12. $\begin{array}{r} 12 \\ \times\ 5 \\ \hline \end{array}$

13. $\begin{array}{r} 11 \\ \times\ 7 \\ \hline \end{array}$

14. $\begin{array}{r} 9 \\ \times\ 8 \\ \hline \end{array}$

15. $\begin{array}{r} 6 \\ \times\ 6 \\ \hline \end{array}$

16. $\begin{array}{r} 12 \\ \times\ 8 \\ \hline \end{array}$

17. $5\overline{)85}$

18. $9\overline{)81}$

19. $4\overline{)88}$

20. $12\overline{)144}$

21. $7\overline{)56}$

Equivalent Fractions

Vocabulary

Fill in the blank.

1. A fraction whose numerator and denominator can both be divided evenly only by 1 is in _____.

Write two equivalent fractions for each.

2. $\frac{5}{10}$ _____

3. $\frac{6}{18}$ _____

4. $\frac{3}{6}$ _____

5. $\frac{8}{20}$ _____

6. $\frac{4}{12}$ _____

7. $\frac{10}{20}$ _____

8. $\frac{1}{4}$ _____

9. $\frac{9}{36}$ _____

Tell whether each fraction is in simplest form. If not, write it in simplest form.

10. $\frac{3}{4}$ _____

11. $\frac{3}{6}$ _____

12. $\frac{4}{5}$ _____

13. $\frac{3}{7}$ _____

14. $\frac{9}{12}$ _____

15. $\frac{2}{8}$ _____

16. $\frac{16}{32}$ _____

17. $\frac{3}{5}$ _____

Find the missing numerator or denominator.

18. $\frac{6}{12} = \frac{1}{2}$

19. $\frac{3}{9} = \frac{1}{1}$

20. $\frac{3}{12} = \frac{1}{1}$

21. $\frac{5}{15} = \frac{1}{3}$

22. $\frac{4}{10} = \frac{2}{1}$

23. $\frac{9}{18} = \frac{1}{2}$

24. $\frac{4}{16} = \frac{1}{1}$

25. $\frac{12}{24} = \frac{1}{2}$

Mixed Review

Estimate.

26. $6,834 \times 28$ _____

27. $975 \div 11$ _____

28. $3,210 \times 49$ _____

29. $495 \div 52$ _____

30. $888 \div 29$ _____

31. $9,011 \times 11$ _____

Name _____

Compare and Order Fractions

Compare the fractions. Write <, >, or = in the ○.

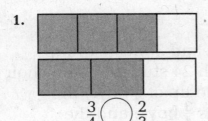

1.
$\frac{3}{4}$ ○ $\frac{2}{3}$

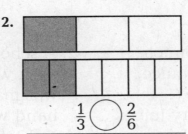

2.
$\frac{1}{3}$ ○ $\frac{2}{6}$

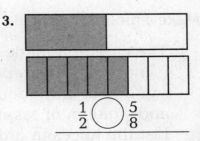

3.
$\frac{1}{2}$ ○ $\frac{5}{8}$

Write <, >, or = in each ○.

4. $\frac{1}{3}$ ○ $\frac{1}{4}$

5. $\frac{5}{6}$ ○ $\frac{4}{6}$

6. $\frac{1}{2}$ ○ $\frac{6}{12}$

7. $\frac{3}{4}$ ○ $\frac{3}{5}$

8. $\frac{2}{5}$ ○ $\frac{3}{5}$

9. $\frac{1}{8}$ ○ $\frac{1}{7}$

10. $\frac{2}{4}$ ○ $\frac{1}{2}$

11. $\frac{4}{8}$ ○ $\frac{4}{10}$

Order the fractions from *greatest* to *least*.

12. $\frac{2}{5}, \frac{1}{5}, \frac{3}{5}$

13. $\frac{2}{6}, \frac{1}{4}, \frac{2}{5}$

14. $\frac{1}{6}, \frac{1}{3}, \frac{1}{2}$

15. $\frac{3}{4}, \frac{2}{3}, \frac{5}{8}$

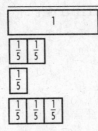

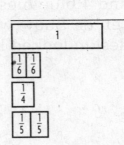

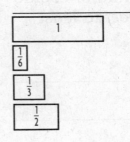

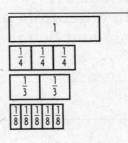

Order the fractions from *least* to *greatest*.

16. $\frac{3}{12}, \frac{4}{10}, \frac{2}{3}$

17. $\frac{5}{8}, \frac{1}{2}, \frac{2}{3}$

18. $\frac{1}{4}, \frac{1}{6}, \frac{1}{5}$

19. $\frac{4}{6}, \frac{7}{12}, \frac{2}{5}$

Mixed Review

Write each fraction in simplest form.

20. $\frac{3}{12}$ _____

21. $\frac{5}{25}$ _____

22. $\frac{6}{18}$ _____

23. $\frac{7}{49}$ _____

Add or multiply.

24. $\begin{array}{r} 7,919 \\ \times \quad 4 \\ \hline \end{array}$

25. $\begin{array}{r} 4,111 \\ + \quad 16 \\ \hline \end{array}$

26. $\begin{array}{r} 3,219 \\ + 1,808 \\ \hline \end{array}$

27. $\begin{array}{r} 6,425 \\ \times \quad 9 \\ \hline \end{array}$

Problem Solving Strategy

Make a Model

Make a model to solve.

1. The cafeteria made a punch using $\frac{1}{2}$ gallon of apple juice, $\frac{5}{8}$ gallon of orange juice, and $\frac{2}{3}$ gallon of raspberry juice. List the juices in order from *greatest* to *least* amount used.

   _____

2. A school had 3 music groups, each with 24 students. The choir was made up of $\frac{1}{3}$ boys, the band was $\frac{3}{4}$ boys, and the orchestra was $\frac{5}{8}$ boys. Which music group had the greatest fraction of girls?

   _____

3. Matt bought cookies at a bakery. He bought $\frac{1}{2}$ dozen oatmeal cookies, $\frac{2}{3}$ dozen cinnamon cookies, and $\frac{3}{4}$ dozen chocolate cookies. List each kind of cookie in order from *greatest* to *least* amount bought.

   _____

   _____

4. Katrina made a square design with 25 tiles. She used 9 red tiles for the diagonals, 12 yellow tiles to complete the outside border, and 4 blue tiles to complete the center. Show what Katrina's design looked like.

Mixed Review

5. $13\overline{)6,249}$

6. $8\overline{)9,122}$

7. $12\overline{)2,424}$

8. $4\overline{)3,175}$

Find the value of the expression.

9. $12 \times (9 - 3) =$ _____

10. $(4 + 4) \times 8 =$ _____

11. $(15 - 4) \times 9 =$ _____

Name _____

Mixed Numbers

Vocabulary

Fill in the blank.

1. A _____ is made up of
   a whole number and a fraction.

Write a mixed number for each picture.

2.

3.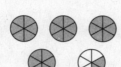

4.

_____   _____   _____

Rename each fraction as a mixed number and each mixed number as a
fraction. You may wish to draw a picture.

5. $\dfrac{16}{3}$ _____   6. $\dfrac{9}{2}$ _____   7. $2\dfrac{5}{6}$ _____   8. $3\dfrac{1}{4}$ _____

For Exercises 9–11, use the figures at the right.

9. How many whole figures are shaded?
   Write an expression for the shaded part
   in the last figure.

   _____

10. How can you change the model to show 5 wholes?

    _____

11. What fraction and mixed number can you write for

    the shaded parts of the figures? _____

Mixed Review

12. $4 \times 4 =$ ___   13. $9 \times 5 =$ ___   14. $8 \times 7 =$ ___   15. $24 \times 1 =$ ___

16. $48 \div 12 =$ ___   17. $88 \div 11 =$ ___   18. $72 \div 9 =$ ___   19. $121 \div 11 =$ ___

Name _____

Add Like Fractions

Find the sum.

1. $\frac{3}{6} + \frac{1}{6} =$ ____

2. $\frac{1}{8} + \frac{6}{8} =$ ____

3. $\frac{3}{5} + \frac{4}{5} =$ ____

4. $\frac{5}{12} + \frac{2}{12} =$ ____

5. $\frac{6}{10} + \frac{7}{10} =$ ____

6. $\frac{3}{4} + \frac{2}{4} =$ ____

7. $\begin{array}{r} \frac{2}{5} \\ + \frac{1}{5} \\ \hline \end{array}$

8. $\begin{array}{r} \frac{5}{9} \\ + \frac{4}{9} \\ \hline \end{array}$

9. $\begin{array}{r} \frac{2}{11} \\ + \frac{4}{11} \\ \hline \end{array}$

Compare. Write $<$, $>$, or $=$ in each $\bigcirc$.

10. $\frac{2}{9} + \frac{3}{9}$ $\bigcirc$ $\frac{4}{9}$

11. $\frac{1}{6} + \frac{2}{6}$ $\bigcirc$ $\frac{1}{2}$

12. $\frac{5}{9} + \frac{8}{9}$ $\bigcirc$ 1

Find the value of n.

13. $\frac{2}{7} + \frac{4}{n} = \frac{6}{7}$ _____

14. $\frac{3}{13} + \frac{n}{13} = \frac{9}{13}$ _____

15. $\frac{6}{9} + \frac{1}{n} = \frac{7}{9}$ _____

16. $\frac{9}{n} + \frac{3}{12} = 1$ _____

Mixed Review

17. $7 + 7 + 7 + 7 =$ _____

18. $12 + 12 + 12 + 12 + 12 =$ _____

19. $\begin{array}{r} 8 \\ \times 7 \\ \hline \end{array}$

20. $\begin{array}{r} 10 \\ \times 5 \\ \hline \end{array}$

21. $\begin{array}{r} 3 \\ \times 9 \\ \hline \end{array}$

22. $\begin{array}{r} 7 \\ \times 7 \\ \hline \end{array}$

23. $\begin{array}{r} 6 \\ \times 9 \\ \hline \end{array}$

Write an equivalent fraction for each.

24. $\frac{7}{14} =$ ____

25. $\frac{16}{40} =$ ____

26. $\frac{12}{36} =$ ____

27. $\frac{9}{90} =$ ____

28. $\frac{6}{18} =$ ____

Subtract Like Fractions

Use fraction bars or draw a picture to find the difference.

1. $\frac{3}{4} - \frac{2}{4} =$ _____

2. $\frac{4}{6} - \frac{3}{6} =$ _____

3. $\frac{7}{8} - \frac{3}{8} =$ _____

4. $\frac{5}{10} - \frac{3}{10} =$ _____

5. $\frac{3}{5} - \frac{1}{5} =$ _____

6. $\frac{6}{8} - \frac{2}{8} =$ _____

7. $\frac{10}{12} - \frac{5}{12} =$ _____

8. $\frac{7}{10} - \frac{3}{10} =$ _____

9. $\frac{5}{6} - \frac{1}{6} =$ _____

Find the difference.

10.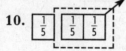

11.

12.

13.

_____   _____   _____   _____

Mixed Review

Find the sum.

14. $\frac{1}{12} + \frac{5}{12} =$ _____

15. $\frac{3}{8} + \frac{3}{8} =$ _____

16. $\frac{4}{7} + \frac{5}{7} =$ _____

17. $\begin{array}{r} 487 \\ \times\ 22 \\ \hline \end{array}$

18. $\begin{array}{r} 68 \\ \times\ 95 \\ \hline \end{array}$

19. $\begin{array}{r} 3{,}287 \\ \times\ 17 \\ \hline \end{array}$

20. $\begin{array}{r} 8{,}061 \\ \times\ 40 \\ \hline \end{array}$

21. $15\overline{)30}$

22. $5\overline{)30}$

23. $3\overline{)30}$

24. $4\overline{)36}$

Name _____

Add and Subtract Mixed Numbers

Find the sum or difference.

1. $5\frac{7}{8}$
$-2\frac{3}{8}$

2. $6\frac{4}{10}$
$+4\frac{3}{10}$

3. $9\frac{3}{4}$
$+2\frac{2}{4}$

4. $3\frac{2}{3}$
$-2\frac{1}{3}$

5. $5\frac{4}{5}$
$+1\frac{2}{5}$

6. $8\frac{6}{8}$
$-3\frac{2}{8}$

7. $9\frac{8}{12}$
$+6\frac{4}{12}$

8. $4\frac{5}{6}$
$-3\frac{3}{6}$

9. $7\frac{8}{9}$
$-6\frac{1}{9}$

10. $9\frac{9}{10}$
$+5\frac{2}{10}$

11. $8\frac{2}{4}$
$+6\frac{1}{4}$

12. $3\frac{10}{12}$
$-1\frac{7}{12}$

13. $7\frac{4}{5} - 1\frac{3}{5} =$ _____

14. $9\frac{5}{8} + 4\frac{4}{8} =$ _____

15. $4\frac{6}{9} - 2\frac{2}{9} =$ _____

16. $5\frac{9}{12} + 2\frac{3}{12} =$ _____

17. $9\frac{2}{5} + 3\frac{1}{5} =$ _____

18. $6\frac{7}{10} - 2\frac{5}{10} =$ _____

Compare. Write $<$, $>$, or $=$ in each $\bigcirc$.

19. $6\frac{1}{7} + 3\frac{5}{7} \bigcirc 10$

20. $3\frac{1}{4} \bigcirc 1\frac{5}{8} + 1\frac{5}{8}$

21. $16\frac{7}{10} - 7\frac{7}{10} \bigcirc 10$

Mixed Review

22. 48
$+ 78$

23. 63
$- 57$

24. 140
$- 79$

25. 224
$+ 865$

26. 370
$- 263$

27. 586
$- 139$

28. 428
$+ 765$

29. 831
$- 156$

30. 605
$- 384$

31. 372
$- 189$

© Harcourt

Problem Solving Skill

Choose the Operation

Write the operation. Then solve each problem.

1. Henry and Cyndi each ate $\frac{1}{3}$ of a small cake. What fraction of the cake did they eat altogether?

   _____

2. Linda baked a huge cookie for her friends. Sue ate $\frac{5}{8}$ of the cookie and Mary ate $\frac{3}{8}$. How much more of the cookie did Sue eat?

   _____

3. Phillip likes to ride his bike, to skateboard, and to read in his spare time. He spends $\frac{2}{8}$ of his time riding his bike and $\frac{5}{8}$ of his time skateboarding. How much of his spare time does he have left to spend reading?

   _____

4. Mr. Jones baked 12 cupcakes for the class party. Before lunch $\frac{3}{12}$ of the cupcakes were eaten. After lunch $\frac{5}{12}$ of the cupcakes were eaten. What fraction of the cupcakes were left for a snack after school?

   _____

Mixed Review

Solve.

5. At the end of five days, Joseph had saved $30. If each day he saved $2 more than the day before, how much money did Joseph save each day?

   _____

6. A series of numbers starts with 2. Each number in the series is two times as great as the number before it. What is the sixth number in the series?

   _____

7.	$20.22 + $15.24	8.	$38.40 − $19.99	9.	2,649 − 1,670	10.	9,028 + 3,840	11.	$38.20 + $88.79

Add Unlike Fractions

Use fraction bars to find the sum.

1.
1

$\frac{1}{3}$	$\frac{1}{3}$	$\frac{1}{6}$

_____

2.
1

$\frac{1}{4}$	$\frac{1}{4}$	$\frac{1}{8}$	$\frac{1}{8}$	$\frac{1}{8}$

_____

3.
1

$\frac{1}{3}$	$\frac{1}{3}$	$\frac{1}{4}$

_____

4.
1

$\frac{1}{2}$	$\frac{1}{5}$

_____

5.
1

$\frac{1}{12}$	$\frac{1}{12}$	$\frac{1}{12}$	$\frac{1}{3}$

_____

6.
1

$\frac{1}{10}$	$\frac{1}{10}$	$\frac{1}{10}$	$\frac{1}{5}$

_____

7. $\frac{1}{3} + \frac{1}{6}$

_____

8. $\frac{5}{8} + \frac{3}{4}$

_____

9. $\frac{3}{4} + \frac{1}{6}$

_____

10. $\frac{7}{10} + \frac{2}{5}$

_____

11. $\frac{4}{10} + \frac{3}{5}$

_____

12. $\frac{4}{5} + \frac{7}{10}$

_____

Mixed Review

13. $19\overline{)4,999}$ **14.** $32\overline{)6,471}$ **15.** $17\overline{)219}$ **16.** $3\overline{)8,536}$ **17.** $8\overline{)830}$

Name _____

Subtract Unlike Fractions

Use fraction bars to find the difference.

1.
$\frac{1}{2}$		
$\frac{1}{12}$ $\frac{1}{12}$ $\frac{1}{12}$		?

_____

2.
$\frac{1}{3}$	
$\frac{1}{9}$ $\frac{1}{9}$	?

_____

3.
$\frac{1}{4}$	$\frac{1}{4}$	$\frac{1}{4}$
$\frac{1}{8}$ $\frac{1}{8}$	?	

_____

4.
$\frac{1}{3}$	$\frac{1}{3}$
$\frac{1}{12}$ $\frac{1}{12}$ $\frac{1}{12}$ $\frac{1}{12}$ $\frac{1}{12}$	?

_____

5.
$\frac{1}{10}$ $\frac{1}{10}$ $\frac{1}{10}$ $\frac{1}{10}$ $\frac{1}{10}$ $\frac{1}{10}$ $\frac{1}{10}$			
$\frac{1}{5}$	$\frac{1}{5}$	$\frac{1}{5}$	?

_____

6.
$\frac{1}{12}$ $\frac{1}{12}$ $\frac{1}{12}$ $\frac{1}{12}$ $\frac{1}{12}$ $\frac{1}{12}$ $\frac{1}{12}$ $\frac{1}{12}$ $\frac{1}{12}$ $\frac{1}{12}$			
$\frac{1}{4}$	$\frac{1}{4}$	$\frac{1}{4}$	?

_____

7. $\frac{4}{5} - \frac{3}{10}$

_____

8. $\frac{4}{6} - \frac{5}{12}$

_____

9. $\frac{5}{6} - \frac{5}{12}$

_____

10. $\frac{1}{2} - \frac{4}{10}$

_____

11. $\frac{6}{8} - \frac{1}{2}$

_____

12. $\frac{2}{3} - \frac{3}{6}$

_____

13. $\frac{1}{2} - \frac{1}{8}$

_____

14. $\frac{9}{12} - \frac{2}{3}$

_____

15. $\frac{4}{6} - \frac{1}{12}$

_____

16. $\frac{7}{8} - \frac{1}{4}$

_____

17. $\frac{11}{12} - \frac{1}{3}$

_____

18. $\frac{4}{6} - \frac{1}{2}$

_____

Mixed Review

Order from *least* to *greatest*.

19. $\frac{7}{10}, \frac{5}{10}, \frac{2}{5}, \frac{8}{10}$

_____

20. $1\frac{1}{3}, \frac{6}{3}, \frac{1}{6}, \frac{5}{6}$

_____

21. $1, \frac{4}{10}, \frac{8}{10}, \frac{11}{10}$

_____

Record Outcomes

For 1–4, use the table.

Don and Carol organized their outcomes in this table. They used the 3-letter spinner and the 4-number spinner shown.

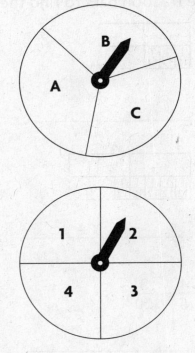

Number	Letter		
	A	B	C
1	II		I
2		III	
3	I		I
4	IIII		II

1. Name all the possible outcomes for this experiment.

_____

_____

2. How many possible outcomes are there?

_____

3. How many outcomes would there be if they had used a 4-letter spinner?

_____

4. In Don and Carol's experiment, which outcome occurred most often?

_____

Mixed Review

5.   318,849
   + 984,741

6.   52,842
   ×   6

7. 17)‾893‾

8. $\frac{5}{12} - \frac{1}{4} =$ _____

9. $\frac{7}{15} - \frac{9}{30} =$ _____

10.   2,875
    +   789

11.   7,932
    − 4,298

12. 14)‾493‾

© Harcourt

Predict Outcomes of Experiments

Write *likely*, *unlikely*, or *equally likely* for the events.

1. tossing an even number or tossing an odd number using a cube numbered 1–6

   _____

2. tossing a prime number on a cube labeled 3, 5, 7, 9, 11, and 13

   _____

3. pulling a yellow marble from a bag with 10 green marbles, 6 red marbles, and 1 yellow marble

   _____

4. the pointer of a spinner with the numbers 1, 2, 3, 3, 3, 3, 3, 3, 6, 6 stopping on 3

   _____

For 5–8, look at the pictures.

5. Which 2 types of marbles are equally likely to be pulled from the bag of marbles?

   _____

6. Which type are you most likely to pull? Why?

   _____

   _____

7. Is it *certain* or *impossible* that the pointer on the spinner will stop on a capital letter?

   _____

8. Is it *certain* or *impossible* that the pointer on the spinner will stop on an M?

   _____

Mixed Review

9. What is the missing number in the pattern?

   2, 3, ___ , 7, 11

10. Estimate the product of 68 and 21.

    _____

Probability as a Fraction

Look at the spinner at the right. Find the probability of each event.

1. the letter _C_ _____

2. the letter _E_ _____

3. a vowel _____

4. a letter in the word _CAB_ _____

5. the letter _F_ _____

6. a consonant _____

7. the letter _A_ _____

Look at the box of marbles. Write _impossible, unlikely, likely,_
or _certain_ for each event, and find the probability.

8. a marble that is not red _____

9. an orange marble _____

10. a green marble _____

11. a red marble _____

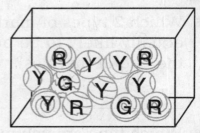

12. a marble that is not green _____

Mixed Review

13. Amanda has a $3\frac{2}{5}$-in. piece of
string and a $2\frac{1}{5}$-in. piece of
string. How much string does
she have?

14. Add. $3\frac{1}{2} + 4\frac{2}{3}$

_____

15. Write <, >, or = in the ◯.

$(42 + 7) - 33$ ◯ $(64 ÷ 8) + 7$

16. Order from _least_ to _greatest._

1.34, 1.32, 0.134, 13.2, 1

_____

More About Probability

For 1–4, use the spinner and the table.

SPINNER EXPERIMENT—100 SPINS			
Outcome	W	X	Y	Z
Tally	⊮⊮ ⊮⊮ ⊮⊮ ⊮⊮ //	⊮⊮ ⊮⊮ ⊮⊮ ⊮⊮ ⊮⊮ //	⊮⊮ ⊮⊮ ⊮⊮ ⊮⊮ /	⊮⊮ ⊮⊮ ⊮⊮ ⊮⊮ ⊮⊮ ⊮⊮

1. What is the mathematical probability of the pointer stopping on each letter on the spinner?

   W _____   Y _____

   X _____   Z _____

2. Use the data in the table. Find the probability of the pointer stopping on each letter in the experiment.

   W _____   Y _____

   X _____   Z _____

3. Use the table to find the probability of the pointer stopping on *W* in the experiment. How does this compare to the mathematical probability?

   _____

4. Compare the probability in the experiment with the mathematical probability of the pointer stopping on X, Y, and Z.

   _____

   _____

   _____

Mixed Review

5. How many possible outcomes are there when you toss a coin and spin the pointer on a spinner with 5 colors?

   _____

6. What kind of triangle is shown below?

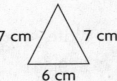

7 cm   7 cm
   6 cm

   _____

Combinations

For 1–6, choose one of each. Find the number of possible combinations by making a tree diagram.

1. Higgins the clown has 3 hats (red, yellow, or blue) to choose from to match his 6 suits (gold, orange, blue, green, purple, and yellow). How many choices does he have?

_____

2. Kathy has 6 different sweaters to wear with her 4 pairs of slacks. How many possible choices does she have?

_____

3. Julia has a choice of using iceberg lettuce or red leaf lettuce for her birthday dinner. In addition, she can choose Italian, Russian, or French salad dressing. How many different outcomes are possible?

_____

4. Thomas had 8 different choices of hats and coats. How many hats might he have? How many coats might he have?

_____

_____

5. Footwear choices:
   Shoes: navy, black, or brown
   Socks: white, black, or tan

_____

6. Event choices:
   Events: sports, play, or movie
   Day: Saturday or Sunday

_____

Mixed Review

7. $(2 \times 4) + (2 \times 2) = $ ▨

_____

8. Round 278,150 to the nearest thousand.

_____

9. Compare. Write $<$, $>$, or $=$.

   379,560 ● 379,561

_____

10. Solve for n.

    $20 - (12 - 2) = n$

_____

Problem Solving Strategy

Make an Organized List

Make an organized list to solve.

1. A spinner is labeled 6, 7, and 8. List all of the possible outcomes of spinning the pointer on the spinner 2 times.

2. Jeanne is writing a report on the computer. She has a choice of 5 different designs for the cover, and 3 different fonts for the report. How many possible ways of writing this report are there?

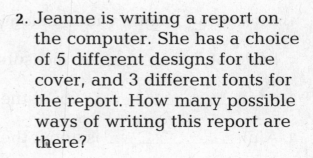

For 3–6, find the possible outcomes of spinning each pointer one time.

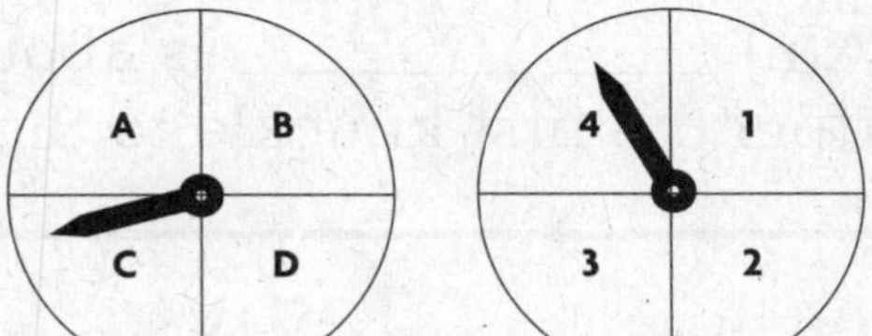

3. How many possible outcomes are there?

4. List all of the possible outcomes.

5. How many possible outcomes would there be if the numbered spinner had 6 numbers?

6. How many of the possible outcomes include the letter F?

Mixed Review

7. The race started at 6:53 P.M. and ended at 7:14 P.M. How long did the race take?

8. $\begin{array}{r} 67{,}314 \\ \times \quad 6 \\ \hline \end{array}$

9. $(6 \times 4) - (3 \times 2) = \blacksquare$

10. Round 4,278,555 to the nearest ten thousand.

Length: Choose the Appropriate Unit

Vocabulary

Complete.

1. Measuring length, width, height, and distance are all

   forms of _____ measurement.

2. A(n) _____ is about the length of a baseball bat.

3. A(n) _____ is about the distance you can walk in 20 minutes.

4. A(n) _____ is about the length of a sheet of paper.

5. A(n) _____ is about the length of your thumb
   from the first knuckle to the tip.

Choose the most reasonable unit of measure. Write *in.*, *ft*, *yd*, or *mi*.

6. The length of a calculator is about 4 _____.

7. The height of a flagpole is about 25 _____.

8. The height of a refrigerator is about 2 _____.

9. The distance along the walkathon is 12 _____.

Name the greater measurement.

10. 50 ft or 50 yd

11. 17 mi or 17 yd

12. 243 in. or 243 yd

_____

_____

_____

Mixed Review

13. $\frac{1}{6} + \frac{2}{3}$

14. $\frac{5}{6} + \frac{2}{3}$

15. Write $\frac{10}{15}$ as a fraction
    in simplest form.

_____

_____

_____

© Harcourt

Measure Fractional Parts

Estimate to the nearest inch. Then measure to the nearest $\frac{1}{4}$ inch.

1.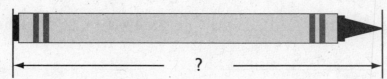

_____

Estimate to the nearest $\frac{1}{2}$ inch. Then measure to the nearest $\frac{1}{8}$ inch.

2.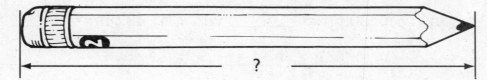

_____

Order the measurements from *least* to *greatest*.

3. $4\frac{1}{8}$ in.; $3\frac{1}{2}$ in.; $4\frac{1}{4}$ in.; $4\frac{3}{8}$ in.

4. $\frac{1}{8}$ in.; $\frac{1}{2}$ in.; $\frac{3}{4}$ in.; $\frac{5}{8}$ in.

_____

Mixed Review

For 5–6, use the Tree Growth Chart.

5. To the nearest foot, how tall was the tree in the first year? second year? third year? fourth year?

_____

6. Between which two years did the tree grow the most?

_____

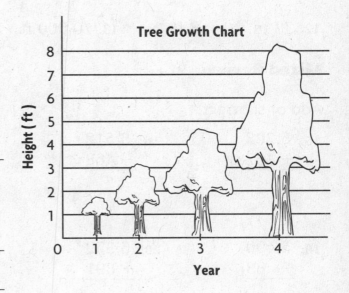

Algebra: Change Linear Units

Complete. Tell whether you multiply or divide.

1. 48 in. = _____ ft   2. 36 ft = _____ yd   3. 4 yd = _____ in.

_____   _____   _____

4. 3 mi = _____ ft   5. 2 mi = _____ yd   6. 5 mi = _____ ft

_____   _____   _____

7. 7 ft = _____ in.   8. 300 ft = _____ yd   9. 432 in. = _____ yd

_____   _____   _____

Write an equation that can be used to complete each table.
Complete the table.

10.

Feet, f	3			12	
Yards, y	1	2	3	4	5

_____

11.

Yards, y	1,760		
Miles, m	1	3	4

_____

Compare. Write <, >, or = in the ◯.

12. 38 in. ◯ 3 ft   13. 10,000 ft ◯ 4 mi   14. 100 in ◯ 3 yd

Mixed Review

Add or subtract.

15. 5,283
    + 467

16. 3,512
    − 468

17. 7,536
    − 207

18. 4,106
    − 314

19. 5,490
    − 83

20. 6,372
    + 891

21. 7,536
    + 18

22. 2,013
    − 5

Capacity

Vocabulary

Complete.

1. _____ is the amount a container can hold when filled.

2. Write the word *cup*, *pint*, *quart*, or *gallon* to label each container.

_____   _____   _____   _____

Complete the tables. Change the units.

3.

Cup	Pint
4	
8	
	8

4.

Pint	Quart
4	
	3
8	

5.

Teaspoon	Tablespoon
	1
9	
	6

Choose the most reasonable unit of capacity. Write *cup*, *pint*, *quart*, or *gallon*.

6.

_____

7.

_____

8.

_____

Mixed Review

Round the number to the greatest place value.

9. 3,654 _____   10. 4,399 _____   11. 2,543 _____

12. 17,536 _____   13. 213,502 _____   14. 109,563 _____

Name _____

Weight

Vocabulary

Complete.

1. A bread truck weighs about 2 _____.

2. A slice of bread weighs about 1 _____.

3. A loaf of bread weighs about 1 _____.

Circle the more reasonable measurement.

4. 1,200 lb or 1,200 oz

5. 10 T or 10 lb

6. 68 oz or 68 lb

Complete. Tell whether you multiply or divide.

7. 2 lb = _____ oz

8. 4 T = _____ lb

9. 60,000 lb = _____ T

_____ _____ _____

10. 48 oz = _____ lb

11. 1 T = _____ oz

12. 208 oz = _____ lb

_____ _____ _____

Change to pounds.

13. 64 oz = _____ lb

14. 6 T = _____ lb

15. 128 oz = _____ lb

Mixed Review

Write the product or quotient.

16. $6 \times 3 =$ _____

17. $10 \times 3 =$ _____

18. $6 \times 5 =$ _____

19. $35 \div 7 =$ _____

20. $7 \times 6 =$ _____

21. $18 \div 3 =$ _____

22. $7 \times 8 =$ _____

23. $36 \div 6 =$ _____

24. $8 \times 11 =$ _____

Name _____

Problem Solving Strategy

Compare Strategies

Choose a strategy to solve.

1. Sarah is making a large pot of soup. She adds 7 quarts of water and 3 pints of tomato juice. What is the least number of one-pint servings she can make?

_____

2. Along the 30-foot wall, there is a plant every 6 feet. The plants start at one end of the wall. How many plants are there?

_____

3. Roland is buying sod for some patches on his lawn. Each patch needs 4 feet of sod. He buys 5 yards of sod. How many patches can he cover?

_____

4. Karla is making tea for some friends. Each cup of tea uses 1 cup of water. Karla fills a 3-quart pitcher with water. How many teacups can she fill?

_____

5. Cherie's town is bagging aluminum cans for recycling. Each bag holds 5 pounds of cans. They need to collect 2 tons of cans before their donation will be accepted. How many bags of cans will they need?

_____

6. Henry collected 10 cans in the first hour, 15 cans the second hour, and 20 cans the third hour. If this pattern continues, how many cans will he collect in all in six hours?

_____

Mixed Review

7. $\begin{array}{r} 314 \\ \times \ \ 4 \\ \hline \end{array}$

8. $\begin{array}{r} 236 \\ \times \ \ 3 \\ \hline \end{array}$

9. $\begin{array}{r} 413 \\ + \ 37 \\ \hline \end{array}$

10. $\begin{array}{r} 207 \\ \times \ \ 4 \\ \hline \end{array}$

11. $\begin{array}{r} 535 \\ + 493 \\ \hline \end{array}$

12. $\begin{array}{r} 537 \\ + 395 \\ \hline \end{array}$

13. $\begin{array}{r} 537 \\ \times \ \ 5 \\ \hline \end{array}$

14. $\begin{array}{r} 716 \\ + 239 \\ \hline \end{array}$

15. $\begin{array}{r} 716 \\ \times \ \ 9 \\ \hline \end{array}$

16. $\begin{array}{r} 375 \\ + 909 \\ \hline \end{array}$

Metric Length

Vocabulary

Complete.

1. A _____ is about the width of your index finger.

2. A _____ is about the width of an adult's hand.

3. A _____ is about the thickness of a dime.

4. A _____ is about the length of 10 football fields.

Use a centimeter ruler or a meterstick to measure each item.
Write the measurement and unit of measure used.

5. length of your desk   6. width of a piece of chalk   7. height of a tree

_____   _____   _____

Choose the most reasonable unit of measure. Write *a*, *b*, or *c*.

8. _____ width of a pen   a. 8 cm   b. 8 mm   c. 8 dm

9. _____ distance around   a. 1,000 cm   b. 1,000 km   c. 1,000 m
           the school

10. _____ height of a tree   a. 5 km   b. 5 dm   c. 5 m

11. _____ distance between   a. 22 km   b. 22 dm   c. 22 m
           two towns

Mixed Review

12. 15   13. 1,000   14. 17   15. 382
    × 10   ×   12   × 43   ×   6

16. 350 × n = 35,000   n = _____   17. n × 36 = 360   n = _____

Algebra: Change Linear Units

Complete. Tell whether you multiply or divide by 10, 100, or 1,000.

1. 300 cm = _____ m   **2.** 3 km = _____ m   **3.** 4,000 m = _____ km

_____   _____   _____

4. 50 m = _____ dm   **5.** 40 km = _____ m   **6.** 68 m = _____ cm

_____   _____   _____

Write the correct unit.

7. 500 cm = 5 _____   **8.** 60 dm = 6 _____   **9.** _____ km = 8,000 m

10. 20 cm = 2 _____   **11.** 3,000 m = 3 _____   **12.** 200 m = 20,000 _____

Compare. Write <, >, or = in each ◯.

13. 12 m ◯ 120 cm   **14.** 14 m ◯ 1,400 cm   **15.** 3 km ◯ 4,000 m

16. 4 m ◯ 3 km   **17.** 30 dm ◯ 3 m   **18.** 300 m ◯ 3,000 dm

Order from least to greatest.

19. 2 m; 100 cm; 4 dm; 3 km   **20.** 3,000 m; 3 dm; 300 km; 3,000 cm

_____   _____

Mixed Review

21. Which customary unit of length would be best to give the distance across a soccer field?

_____

22. Write an expression for 3 times the number of people, p, at the county fair.

_____

23. 84
× 62

24. 48,588
− 40,315

25. 315
× 27

26. $4\overline{)3,788}$

27. $6\overline{)973}$ **28.** $8\overline{)5,800}$ **29.** $12\overline{)144}$ **30.** $4\overline{)3,604}$

Capacity

Vocabulary

Complete.

1. A _____ is about the size of a sports-drink bottle.
   It contains 1,000 milliliters.

2. A _____ is about the size of a drop of liquid in an eyedropper.

Choose the more reasonable unit of measure. Write *mL* or *L*.

3. wading pool

4. a soda can

5. a baby bottle

_____ _____ _____

Choose the best estimate. Circle *a, b,* or *c.*

6.
   a. 3 mL
   b. 30 mL
   c. 3 L

7.
   a. 42 mL
   b. 420 mL
   c. 42 L

8.
   a. 62 mL
   b. 620 mL
   c. 62 L

Complete.

9. 5 L = _____ mL

10. 70 L = _____ mL

11. 4 L = _____ mL

12. 6,000 mL = _____ L

13. 12,000 mL = _____ L

14. 26,000 mL = _____ L

Mixed Review

15. $\begin{array}{r} 187 \\ +435 \\ \hline \end{array}$

16. $\begin{array}{r} 461 \\ \times\ 34 \\ \hline \end{array}$

17. $4\overline{)723}$

18. $5\frac{3}{16} + 7\frac{1}{16}$

19. Ron's car has a 12-gallon gas tank. If gas costs $1.45 per gallon, how much will it cost to fill the tank?

20. A 5-lb bag of flour costs $1.10. A 20-oz bag of flour costs $0.40. Which is the better buy?

_____ _____

21. 5 km = _____ m

22. 71 m = _____ cm

23. 98 m = _____ mm

Name _____

Mass

Vocabulary

Write the letter of the word that is best described.

1. _____ the amount of mass that is about equal to a baseball bat

2. _____ the amount of matter in an object

3. _____ the amount of mass that is about equal to a small paper clip

a. kilogram (kg)

b. gram (g)

c. mass

Choose the more reasonable measurement.

4.	5.	6.	7.

1 g or 1 kg	5 g or 5 kg	200 g or 20 kg	600 g or 600 kg
_____	_____	_____	_____

Complete.

8. 3 kg = _____ g   **9.** 14 kg = _____ g   **10.** _____ kg = 20,000 g

Mixed Review

11. One serving of macaroni and cheese is 70 g. How many kilograms are needed to serve 200 people?

_____

12. If 3 servings of macaroni and cheese cost $0.99, how much will it cost to serve 210 people?

_____

13. 72)‾4,216   **14.** 19)‾103   **15.** 10)‾20,000   **16.** 24)‾1,920

© Harcourt

Problem Solving Strategy

Draw a Diagram

Draw a diagram to solve.

1. Steve and Sara bought a total of 14 items at the grocery store. Sara bought two more than twice the number of items that Steve bought. How many items did each buy?

   _____

   _____

   _____

2. Mike, Thea, and Emily were reading library books. Mike read 4 books. Thea read 2 more than twice the number of books that Emily read. Emily read 1 book less than Mike. How many books did each person read?

   _____

   _____

   _____

3. Tina, Kevin, and Amy flew their kites. Kevin's kite flew 2 meters higher than Amy's. Tina's flew 1 meter lower than half as high as Amy's. Amy's kite flew 300 decimeters high. How high did Tina's and Kevin's kites fly?

   _____

   _____

   _____

4. Jim's family went hiking. Jim was able to hike 5 miles. His mother and father each hiked 1 mile more than three times the distance that Jim hiked. Jim's brother Tim hiked 1 mile less than Jim did. How far did each person hike?

   _____

   _____

   _____

Mixed Review

5. 300 m = _____ cm

6. 400 dm = _____ m

7. 7,000 m = _____ km

8. 20 ft = _____ in.

9. 4 lb = _____ oz

10. 1 pt = _____ c

11. 48 in. = _____ ft

12. 6 c = _____ pt

13. 20 qt = _____ gal

Name _____

Relate Benchmark Measurements

Estimate the conversion.

1. 6 kg ≈ _____ lb
2. 3 mi ≈ _____ km
3. 5 cm ≈ _____ in.

4. 2 oz ≈ _____ g
5. 10 qt ≈ _____ L
6. 12 ft ≈ _____ m

7. 4 lb ≈ _____ kg
8. 9 in. ≈ _____ cm
9. 84 g ≈ _____ oz

Compare. Write < or > in each ◯.

10. 2 L ◯ 4 qt
11. 3 m ◯ 7 ft
12. 3 mi ◯ 4 km

13. 2 in. ◯ 8 cm
14. 6 lb ◯ 2 kg
15. 80 g ◯ 3 oz

Mixed Review

16. 68 × 35
17. 52 × 27
18. 81 × 49
19. 77 × 68

20. 435 × 19
21. 212 × 34
22. 408 × 25
23. 822 × 56

24. 14)‾448
25. 18)‾990
26. 22)‾836
27. 27)‾405

28. 16)‾196
29. 23)‾483
30. 14)‾490
31. 31)‾936

© Harcourt

Practice PW137

Relate Fractions and Decimals

Write the decimal and fraction shown by each model or number line.

1. 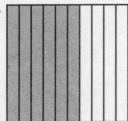   2. 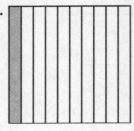   3.    4.

_____   _____   _____   _____

5.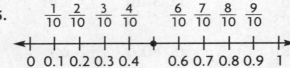

_____

6.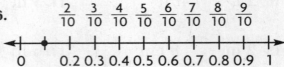

_____

Write each fraction as a decimal. Use a model to help you.

7. $\frac{6}{10}$ _____   8. $\frac{8}{10}$ _____   9. $\frac{39}{100}$ _____   10. $\frac{64}{100}$ _____

Write the decimal two other ways.

11. 0.2 _____   12. 0.4 _____   13. 0.12 _____   14. 0.66 _____

_____   _____   _____   _____

Mixed Review

15. During his vacation, Brian used 7 rolls of 24-photo film. How many photos did Brian take?

16. If 8 students can sit at one table, how many tables are needed to seat 134 students?

_____   _____

17. 17
    × 17

18. 11
    × 11

19. 100
    × 100

20. 18
    × 18

21. 12
    × 12

Decimals to Thousandths

Write each decimal as a fraction.

1. 0.089   **2.** 0.001   **3.** 0.234   **4.** 0.090   **5.** 0.025

_____   _____   _____   _____   _____

6. 0.988   **7.** 0.543   **8.** 0.087   **9.** 0.751   **10.** 0.009

_____   _____   _____   _____   _____

Write each fraction as a decimal.

11. $\frac{12}{1,000}$ **12.** $\frac{285}{1,000}$ **13.** $\frac{9}{1,000}$ **14.** $\frac{364}{1,000}$ **15.** $\frac{70}{1,000}$

_____   _____   _____   _____   _____

Write the decimal two other ways.

16. 0.9 + 0.04 + 0.003   **17.** 0.258   **18.** 0.3 + 0.01 + 0.006

_____   _____   _____

_____   _____   _____

19. twenty-seven
thousandths

20. one hundred
seventy-nine
thousandths

21. 0.064

_____   _____   _____

_____   _____   _____

Mixed Review

Compare. Write <, >, or = in each ◯.

22. 2,431 ◯ 2,043   **23.** 70,450 ◯ 70,450   **24.** 1,382 ◯ 1,823

Solve.

25. 11 × 15   **26.** 98 + 165,424   **27.** 12,089 − 10,078

_____   _____   _____

Equivalent Decimals

Vocabulary

Complete.

1. _____ are decimals that name the same number.

Are the two decimals equivalent? Write *equivalent* or *not equivalent*.

2. 0.4 and 0.40 _____

3. 0.1 and 0.01 _____

4. 0.50 and 0.5 _____

5. 0.20 and 0.02 _____

6. 0.3 and 0.30 _____

7. 0.80 and 0.8 _____

8. 0.9 and 0.90 _____

9. 0.18 and 0.81 _____

Write an equivalent decimal for each. You may use decimal models.

10. 0.7 _____

11. 0.1 _____

12. 0.60 _____

13. 0.4 _____

14. 0.20 _____

15. 0.8 _____

16. 0.30 _____

17. 0.5 _____

18. 0.90 _____

19. 0.3 _____

Mixed Review

20. $\frac{7}{10} + \frac{7}{10} =$ _____

21. $1\frac{4}{5} + 1\frac{4}{5} =$ _____

22. $3\frac{8}{9} + 3\frac{8}{9} =$ _____

23. $5\frac{4}{5} - 1\frac{3}{5} =$ _____

24. $\frac{10}{9} + 3\frac{5}{9} =$ _____

25. $\frac{7}{6} - \frac{2}{3} =$ _____

26. $\frac{4}{7} + \frac{2}{7} =$ _____

27. $1\frac{3}{4} + 2\frac{3}{4} =$ _____

28. $7\frac{2}{3} + 6\frac{1}{3} =$ _____

29. $\frac{3}{4} - \frac{1}{2} =$ _____

30. $6\frac{5}{6} - 1\frac{1}{6} =$ _____

31. $\frac{4}{9} + 7\frac{7}{9} =$ _____

Relate Mixed Numbers and Decimals

Use the number line to write an equivalent mixed number
or decimal for the given letter.

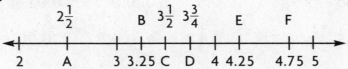

1. A _____   2. B _____   3. C _____

4. D _____   5. E _____   6. F _____

Write a decimal and a mixed number that are equivalent
to each decimal model below.

7.    8.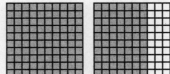

_____   _____

Write an equivalent mixed number or decimal.

9. 12.75 _____   10. 5.50 _____   11. $6\frac{1}{5}$ _____

Mixed Review

12. What digit is in the ten
thousands place in the number
24,639?

_____

13. These are Anna's spelling
scores for 1 week: 86, 90, 85,
94, and 80. What is the median?

_____

14. List the first 5 multiples of 3.

_____

15. List the factors of 50.

_____

Compare and Order Decimals

Compare. Write <, >, or = in each ◯.

1. 0.45 ◯ 0.35   **2.** 0.4 ◯ 0.6   **3.** 0.9 ◯ 0.91   **4.** 0.6 ◯ 0.64

5. 0.50 ◯ 0.55   **6.** 0.7 ◯ 0.17   **7.** 0.02 ◯ 0.22   **8.** 0.49 ◯ 0.4

9. 0.32 ◯ 0.23   **10.** 0.9 ◯ 0.99   **11.** 0.25 ◯ 0.205   **12.** 0.465 ◯ 0.437

Use the number line to order the decimals from *greatest* to *least*.

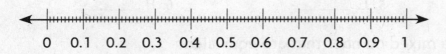

0   0.1   0.2   0.3   0.4   0.5   0.6   0.7   0.8   0.9   1

13. 0.45, 0.54, 0.40, 0.04

_____

14. 0.4, 0.5, 0.04, 0.05, 0.45

_____

15. 0.13, 0.31, 0.3, 0.01, 0.03

_____

16. 0.67, 0.7, 0.76, 0.07, 0.6

_____

17. 0.147, 0.243, 0.202, 0.215, 0.041

_____

18. 0.196, 0.204, 0.13, 0.092, 0.297

_____

Mixed Review

19. Rosie's Umbrella Shop is selling umbrellas for $4.00 off the usual price of $15.00. What is the cost of buying 3 sale umbrellas?

_____

20. To prepare for a presentation, Pete colored $\frac{1}{2}$ of a poster. Rebecca colored $\frac{1}{3}$ of the poster. What fraction still needs to be colored?

_____

Write an equivalent decimal for each.

21. 0.4

_____

22. 0.60

_____

23. 0.8

_____

24. 0.7

_____

Name _____

Problem Solving Strategy

Use Logical Reasoning

Use logical reasoning to solve.

1. Mr. Berg's science class grew tomato plants. The recorded heights of the plants were 13 cm, 15 cm, 17 cm, and 20 cm. Jim's plant was the tallest. Steve's plant was 2 cm taller than Mark's. Eric's plant was the smallest. How tall was Mark's plant?

_____

2. Four students ran a race in gym class. Erica had the fastest time of 10.5 seconds. The other recorded times were 13 seconds, 15 seconds, and 20 seconds. Janie was slower than Erica, but faster than Mike. Joe was the slowest. What were Janie's and Mike's times?

_____

3. Stephanie's class took a spelling test. The scores were 90, 86, 89, 94, and 100. Stephanie got a higher grade than Chris. Sue scored 3 points higher than David. Ellen received the highest score. What was Stephanie's spelling grade?

_____

4. The Nature Club recorded the number of birds at the bird feeder each day for a week. On Monday the club saw 15 birds. The numbers of birds at the feeder on the other days were 12, 13, 19, and 20. On Tuesday, the club saw the fewest birds. On Wednesday, the club saw fewer birds than on Monday. On Friday, the club saw the most birds. How many birds did the club see on Thursday?

_____

Mixed Review

5. $\frac{1}{5} + \frac{2}{5}$

_____

6. List the factors of 21.

_____

7. Order from *least* to *greatest*.
   0.1, 3.00, 0.97, 0.08

_____

8. $128 \div 8$

_____

Round Decimals

Round each to the place of the underlined digit.

1. <u>6</u>.9 _____

2. <u>7</u>.2 _____

3. $8.<u>3</u>2 _____

4. 9.<u>7</u>5 _____

5. 5<u>1</u>.2 _____

6. <u>5</u>.964 _____

7. $84.<u>6</u>5 _____

8. $<u>5</u>.45 _____

Round to the nearest whole number.

9. thirteen and eleven hundredths

_____

10. six and ninety-five hundredths

_____

11. ten and ninety-one hundredths

_____

12. nine and forty-five hundredths

_____

Round to the nearest hundredth.

13. 16.549 _____

14. 31.258 _____

15. 46.953 _____

16. 21.854 _____

17. 25.641 _____

18. 49.397 _____

19. 64.918 _____

20. 87.395 _____

Mixed Review

21. $\begin{array}{r} 429 \\ + 730 \\ \hline \end{array}$

22. $\begin{array}{r} 614 \\ + 88 \\ \hline \end{array}$

23. $\begin{array}{r} 221 \\ + 221 \\ \hline \end{array}$

24. $\begin{array}{r} 4,819 \\ + 2,755 \\ \hline \end{array}$

25. $\begin{array}{r} 1,194 \\ + 3,660 \\ \hline \end{array}$

26. $\begin{array}{r} 879 \\ - 56 \\ \hline \end{array}$

27. $\begin{array}{r} 905 \\ - 548 \\ \hline \end{array}$

28. $\begin{array}{r} 712 \\ - 681 \\ \hline \end{array}$

29. $\begin{array}{r} 3,463 \\ - 2,798 \\ \hline \end{array}$

30. $\begin{array}{r} 5,999 \\ - 590 \\ \hline \end{array}$

31. Solve for n.
$540 \div n = 90$

32. Solve for n.
$(64 - 5) + (12 \div 4) = n$

Estimate Decimal Sums and Differences

Estimate the sum or difference.

1. $\begin{array}{r} 1.5 \\ + 1.2 \end{array}$
2. $\begin{array}{r} 1.8 \\ - 0.6 \end{array}$
3. $\begin{array}{r} 2.3 \\ - 0.7 \end{array}$
4. $\begin{array}{r} 2.94 \\ - 1.13 \end{array}$
5. $\begin{array}{r} 23.94 \\ + 16.98 \end{array}$

6. $\begin{array}{r} 4.25 \\ - 0.86 \end{array}$
7. $\begin{array}{r} 6.45 \\ - 2.63 \end{array}$
8. $\begin{array}{r} \$5.62 \\ + \$2.81 \end{array}$
9. $\begin{array}{r} 16.95 \\ - 3.29 \end{array}$
10. $\begin{array}{r} 45.41 \\ - 29.18 \end{array}$

11. $\begin{array}{r} 1.62 \\ - 1.34 \end{array}$
12. $\begin{array}{r} 3.72 \\ - 1.65 \end{array}$
13. $\begin{array}{r} 2.36 \\ - 1.74 \end{array}$
14. $\begin{array}{r} 3.92 \\ - 1.69 \end{array}$
15. $\begin{array}{r} 3.45 \\ + 2.07 \end{array}$

16. $\begin{array}{r} 23.41 \\ - 11.20 \end{array}$
17. $\begin{array}{r} 2.53 \\ + 1.56 \end{array}$
18. $\begin{array}{r} 3.04 \\ - 1.26 \end{array}$
19. $\begin{array}{r} 2.82 \\ + 2.35 \end{array}$
20. $\begin{array}{r} 4.26 \\ - 2.39 \end{array}$

Mixed Review

Write $<$ or $>$ in each $\bigcirc$.

21. $815 + 37 \bigcirc 850$

22. $1,900 \bigcirc 1,075 + 900$

23. $659 + 659 \bigcirc 1,320$

24. $743 + 643 \bigcirc 1,390$

For 25–26, use the table.

25. If you rounded all of the punt air times to the nearest second, what would be the time that occurred most often?

_____

26. Estimate the difference between Charley's longest time and his shortest time.

| Charley's Football Punt Time in Air ||
Days	Number of Seconds
Monday	3.4 seconds
Tuesday	2.5 seconds
Wednesday	1.7 seconds
Thursday	2.8 seconds
Friday	4.2 seconds

_____

Add Decimals

Write the letter of the model that matches each problem. Solve.

A.   B. 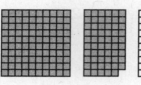   C.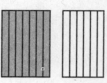

D. 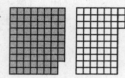   E. 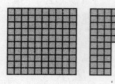   F.

1. $1.35 + 0.64 = n$

_____

2. $0.7 + 0.6 = n$

_____

3. $0.64 + 0.82 = n$

_____

4. $1.59 + 0.43 = n$

_____

5. $0.8 + 0.3 = n$

_____

6. $0.78 + 0.63 = n$

_____

Find the sum. Estimate to check.

7. $\begin{array}{r} 0.6 \\ +\ 0.8 \\ \hline \end{array}$ **8.** $\begin{array}{r} 0.52 \\ +\ 0.39 \\ \hline \end{array}$ **9.** $\begin{array}{r} 0.24 \\ +\ 0.36 \\ \hline \end{array}$ **10.** $\begin{array}{r} 0.593 \\ +0.796 \\ \hline \end{array}$ **11.** $\begin{array}{r} 3.72 \\ +\ 5.88 \\ \hline \end{array}$

12. $\begin{array}{r} 0.9 \\ +\ 0.9 \\ \hline \end{array}$ **13.** $\begin{array}{r} 45.91 \\ +12.57 \\ \hline \end{array}$ **14.** $\begin{array}{r} 0.88 \\ +\ 0.43 \\ \hline \end{array}$ **15.** $\begin{array}{r} 31.504 \\ +14.689 \\ \hline \end{array}$ **16.** $\begin{array}{r} 21.94 \\ +10.28 \\ \hline \end{array}$

Mixed Review

17. Sally bought two packages of hamburger. One package was 2.45 pounds and the other was 3.16 pounds. How many pounds of hamburger did she buy?

_____

18. Henry wanted to buy his friend a treat. He had $3.87. If the treat cost $2.65, about how much money did he have left?

_____

19. $7 \times 7 = $ _____

20. $9 \times 2 = $ _____

21. $4 \times 8 = $ _____

Subtract Decimals

Find the difference. Estimate to check.

1. $\begin{array}{r} 0.9 \\ -\ 0.2 \\ \hline \end{array}$

2. $\begin{array}{r} 0.64 \\ -\ 0.34 \\ \hline \end{array}$

3. $\begin{array}{r} 1.8 \\ -\ 0.3 \\ \hline \end{array}$

4. $\begin{array}{r} 41.526 \\ -32.619 \\ \hline \end{array}$

5. $\begin{array}{r} 1.25 \\ -\ 0.76 \\ \hline \end{array}$

6. $\begin{array}{r} 1.00 \\ -\ 0.56 \\ \hline \end{array}$

7. $\begin{array}{r} 1.62 \\ -\ 0.73 \\ \hline \end{array}$

8. $\begin{array}{r} 17.62 \\ -\ 9.28 \\ \hline \end{array}$

9. $\begin{array}{r} 1.214 \\ -0.478 \\ \hline \end{array}$

10. $\begin{array}{r} 76.43 \\ -34.58 \\ \hline \end{array}$

11. $4.80 - 0.62$

12. $5.99 - 1.03$

13. $20.854 - 11.708$

14. $13.392 - 12.365$

_____   _____   _____   _____

For 15–18, write the missing digits.

15. $4.\underline{\quad} - \underline{\quad}.6 = 2.7$

16. $3\underline{\quad}.5 - \underline{\quad}2.8 = 18.7$

17. $1\underline{\quad}.3 - 8.\underline{\quad} = 6.4$

18. $\underline{\quad}9.2 - \underline{\quad}.4 = 11.8$

Mixed Review

19. What fraction is equivalent to 9.40?

_____

20. Joan's older sister is 1.65 meters tall. Joan is 1.26 meters tall. How much taller is her sister?

_____

21. $\begin{array}{r} 2,875 \\ \times\quad 30 \\ \hline \end{array}$

22. $\begin{array}{r} 7,891 \\ +\ 9,415 \\ \hline \end{array}$

23. $\begin{array}{r} 62,730 \\ -59,881 \\ \hline \end{array}$

24. $\begin{array}{r} 14,962 \\ +29,037 \\ \hline \end{array}$

Add and Subtract Decimals and Money

Find the sum or difference. Estimate to check.

1. $\begin{array}{r} 4.90 \\ + 3.41 \\ \hline \end{array}$ 2. $\begin{array}{r} \$5.20 \\ - \$3.45 \\ \hline \end{array}$ 3. $\begin{array}{r} 5.00 \\ - 2.49 \\ \hline \end{array}$ 4. $\begin{array}{r} 3.50 \\ + 4.62 \\ \hline \end{array}$ 5. $\begin{array}{r} \$35.91 \\ + \$\ 4.00 \\ \hline \end{array}$

6. $\begin{array}{r} 6.90 \\ - 3.81 \\ \hline \end{array}$ 7. $\begin{array}{r} 10 \\ -4.632 \\ \hline \end{array}$ 8. $\begin{array}{r} \$2.60 \\ + \$1.75 \\ \hline \end{array}$ 9. $\begin{array}{r} 5.428 \\ +1.735 \\ \hline \end{array}$ 10. $\begin{array}{r} 7.18 \\ + 2.49 \\ \hline \end{array}$

11. $5.98 − $0.50   12. 35.846 − 4.9   13. 12 − 5.913

_____   _____   _____

Find the missing number.

14. 3.62 − ■ = 1.5   15. 4.96 − 1.2 = ■   16. ■ + 0.29 = 3.81

_____   _____   _____

Mixed Review

17. Sylvia ran 50 meters in 9.62 seconds. Linda finished 0.35 seconds later. Ramie's time was 0.09 seconds more than Linda's. What was Linda's time? Ramie's?

_____

18. Henry bought radish, tomato, and pumpkin seed packages. The radish and tomato seed packages were $0.89 each. The pumpkin seed packages were $1.25 each. How many packages of each kind of seed did he buy if he spent $4.28 in all?

_____

_____

Multiply each number by 72.

19. 4   20. 64   21. 349

_____   _____   _____

Problem Solving Skill

Evaluate Reasonableness of Answers

1. Heidi works as a park ranger giving hiking tours. The trail is 4.3 miles long. If Heidi walks the trail 15 times each week, which is a reasonable estimate of the total number of miles she hikes?

   A Heidi hikes about 100 miles.

   B Heidi hikes about 60 miles.

2. Merrilyn is going to the market to buy produce. She needs 5 pounds of apples at $0.99 per pound and 9 pounds of green beans at $1.29 per pound. Which is a more reasonable estimate of how much money she should bring to the market?

   A $14.00

   B $32.00

For 3–4, use this information.

Peter is reading the instructions on how to build a birdhouse. He needs to cut some pieces of wood from a piece of lumber 100 cm long. The first piece should be 38.9 cm long; the second should be 22.5 cm long.

3. Which is the best estimate for the combined length of the two pieces he cuts?

   A 70 cm   **C** 30 cm

   B 60 cm   **D** 10 cm

4. Which is the best estimate for the length of the remaining lumber after Peter makes the two cuts?

   F 40 cm   **H** 15 cm

   G 20 cm   **J** 10 cm

Mixed Review

5. Find the prime factors of 12. _____

6. List 3 multiples of 10. _____

7. Write the fact family for 3, 5, and 15. _____

8. 90,005
   − 5,842

9. $\frac{9}{10} - \frac{3}{5} =$ _____

10. 52
    × 81

11. 450 ÷ 5 = _____

Name _____

Explore Perimeter

Vocabulary

Complete the sentence.

1. _____ is the distance around a figure.

Find the perimeter of each figure.

2.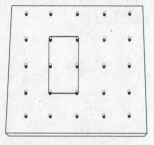

3.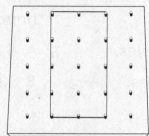

4.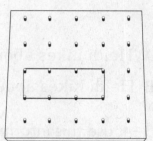

_____   _____   _____

Draw a rectangle with the given perimeter on the dot paper.
Then record the lengths of the sides.

5. 6 units

6. 12 units

_____   _____

Mixed Review

7. $\frac{5}{9} + \frac{2}{9} =$ _____

8. $1\frac{1}{8} + 3\frac{3}{8} =$ _____

9. $\frac{2}{5} + \frac{7}{10} =$ _____

10. $3\frac{4}{5} - 2\frac{2}{5} =$ _____

11. $\frac{5}{6} - \frac{1}{2} =$ _____

12. $3\frac{5}{9} - 1\frac{1}{9} =$ _____

13. $1\frac{1}{2} + 4\frac{1}{8} =$ _____

14. $\frac{3}{4} - \frac{5}{8} =$ _____

15. $2\frac{1}{3} + 1\frac{1}{3} =$ _____

Estimate and Find Perimeter

Vocabulary

Complete the sentence.

1. A _____ is a mathematical rule.

Find the perimeter.

2.

6 in.

4 in.   4 in.

8 in.

_____

3.

5 m   5 m

5 m   5 m

5 m

_____

4.

4 km

4 km   ← 3 km
                    ↓
                    1 km

7 km

_____

5.

6 cm

          5 cm
6 cm
          3 cm
6 cm

_____

6.

5 m
          5 m
10 m
          10 m

          5 cm
15 m

_____

7.

          15 ft
9 ft   8 ft 4 ft
                       6 ft
10 ft   11 ft

_____

Use a formula to find the perimeter.

8.

6 mi

          3 mi

_____

9.

4 ft

          4 ft

_____

10.

14 yd

          5 yd

_____

Mixed Review

11. $\frac{3}{9} + \frac{2}{9} =$ _____

12. $\frac{1}{8} + \frac{5}{8} =$ _____

13. $\frac{9}{10} - \frac{5}{10} =$ _____

14. $\frac{5}{7} - \frac{3}{7} =$ _____

15. $12\overline{)780}$

16. $19\overline{)1,862}$

17. $8\overline{)4,963}$

18. $17\overline{)3,727}$

Circumference

Estimate the circumference.

1.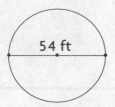
54 ft

_____

2.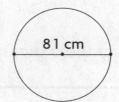
81 cm

_____

3.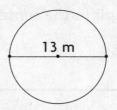
13 m

_____

4.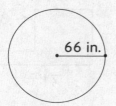
66 in.

_____

5.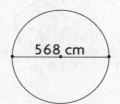
568 cm

_____

6.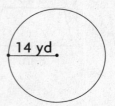
14 yd

_____

7.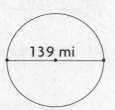
139 mi

_____

8.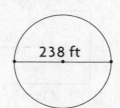
238 ft

_____

9.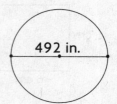
492 in.

_____

10. A wheel has a circumference of 8 inches. It rolls 72 inches. How many complete turns did the wheel make?

_____

Mixed Review

Write the number in word form.

11. 7,849 _____

12. 182 _____

13. 1,283 _____

14. 9,634 _____

15. 17,334 _____

Problem Solving Skill

Use a Formula

Use a formula to solve.

1. Samuel is building a fence to enclose a triangular flower garden. The perimeter of the garden is 7.1 meters. He has completed two sides of the fence, each measuring 2.6 meters. How much of the fence is left to be built?

_____

2. Jessica is putting a ribbon border around the edge of a triangular-shaped flag. The flag's perimeter is 112 cm. If one side of the flag is 35 cm long and another side is 42 cm long, what is the length of the third side?

_____

3. A playground shaped like a triangle has a perimeter of 108 feet. The shortest side is 12 feet long. The longest side is 60 feet long. How long is the missing side?

_____

4. Oliver is walking around a park along a triangular-shaped path. The path's perimeter is 1.75 mi. Oliver has walked around two sides that are 0.45 mi and 0.5 mi long. How far does he have left to walk?

_____

Mixed Review

5. The perimeter of a square window is 72 inches. How long is each side of the window?

_____

6. A rectangular picture 6 inches wide is 4 inches longer than its width. What is the perimeter of the picture?

_____

7. $\begin{array}{r} 891 \\ -\ 493 \end{array}$

8. $\begin{array}{r} 135 \\ -\ 78 \end{array}$

9. $\begin{array}{r} 3,297 \\ -\ 1,368 \end{array}$

10. $\begin{array}{r} 5,204 \\ -\ 675 \end{array}$

11. $\begin{array}{r} 2,005 \\ -\ 1,636 \end{array}$

12. $\frac{7}{9} - \frac{2}{9} =$ _____

13. $\frac{7}{8} - \frac{1}{4} =$ _____

14. $\frac{4}{5} - \frac{1}{10} =$ _____

Estimate Area

Vocabulary

1. _____ is the number of square units needed to cover a surface.

Estimate the area of each figure. Each square is 1 sq yd.

2.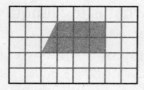

3.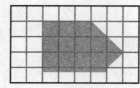

4.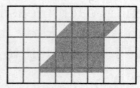

_____

5.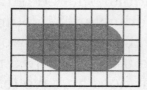

6.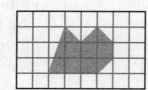

7.

_____

8. Draw a triangle on the grid. Then estimate the area in square units.

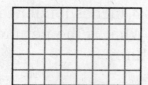

_____

9. Draw a half-circle on the grid. Then estimate the area in square units.

_____

Mixed Review

10.   194
    × 51

11.   428
    × 16

12.   76
    × 24

13.   263
    × 38

14.   93
    × 27

15. 6)973

16. 4)728

17. 14)493

18. 25)475

19. 7)637

Name _____

Find Area

Find the area.

1.

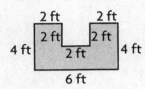

3 yd (top)
3 yd (left) 3 yd (right)
3 yd (bottom)

_____

2.

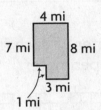

4 cm (top)
1 cm (left) 1 cm (right)
4 cm (bottom)

_____

3.

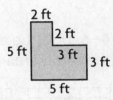

8 in. (top)
5 in. (left) 5 in. (right)
8 in. (bottom)

_____

4.

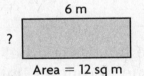

2 ft   2 ft
2 ft   2 ft
4 ft   2 ft   4 ft
6 ft

_____

5.

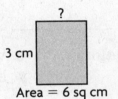

4 mi
7 mi   8 mi
3 mi
1 mi

_____

6.

2 ft
2 ft
5 ft   3 ft   3 ft
5 ft

_____

Find the unknown length.

7.

6 m
?
Area = 12 sq m

_____

8.

?
3 cm
Area = 6 sq cm

_____

9.

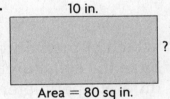

10 in.
?
Area = 80 sq in.

_____

Mixed Review

10. 67
× 16

11. 627
× 41

12. 129
× 76

13. 492
× 10

14. 412
× 89

15. 871
× 13

16. 165
× 64

17. 52
× 37

18. 69
× 28

19. 955
× 31

20. $(7 \times 3) - (4 \times 4) =$ _____

21. $(12 \times 3) - 15 =$ _____

22. $(19 + 28) - (8 \times 2) =$ _____

23. $(17 - 7) + (5 \times 5) =$ _____

Relate Area and Perimeter

For 1–3, find the area and the perimeter of each figure. Then draw another figure that has the same perimeter but a different area.

1.

_____

2.

_____

3.

_____

For 4–6, find the area and perimeter of each figure. Then draw another figure that has the same area but a different perimeter.

4.

_____

5.

_____

6.

_____

7. Which of the figures a–d below have the same area but different perimeters?

_____

8. Which of the figures a–d below have the same perimeter but different areas?

_____

a.

b.

c.

d.

Mixed Review

9. $7\frac{1}{2} + 3\frac{9}{12} =$ _____

10. $4\frac{4}{9} - 1\frac{1}{5} =$ _____

11. $10\frac{6}{7} - 5\frac{2}{14} =$ _____

Circle the prime numbers.

12. 17   33   39   5   142   29   47   30   111   13   52   56   11

Problem Solving Strategy

Find a Pattern

Use *find a pattern* and solve.

1. Alexis is going to put carpet in three rectangular rooms in her house. How do the areas of the rooms change if each room is two times as long and three times as wide as the one before it? Make a table to show how the areas change. Then solve.

   Room 1: l = 4 yd, w = 2 yd

   Room 2: l = 8 yd, w = 6 yd

   Room 3: l = 16 yd, w = 18 yd

2. Douglas has different sizes of rectangular picture frames. How does the perimeter change for each of his picture frames when the width increases by 5 inches? Complete the table and solve.

Picture Frame Sizes			
	Length (in.)	Width (in.)	Perimeter (in.)
Frame A	12	10	
Frame B	12	15	
Frame C	12	20	
Frame D	12	25	

Mixed Review

3. 15 × 7 = _____

4. 121 ÷ 11 = _____

5. 42 × 8 = _____

Name _____

Faces, Edges, and Vertices

Which solid figure do you see in each?

1.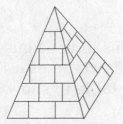

2.

3.

_____   _____   _____

Copy the drawings. Circle each vertex, outline each edge in red, and shade one face in yellow.

4.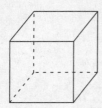

5.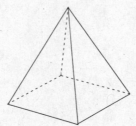

6.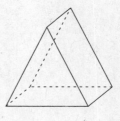

Write the names of the faces and the number of each kind of face of the solid figure.

7. triangular pyramid   8. triangular prism   9. square pyramid

_____   _____   _____

_____   _____   _____

Mixed Review

Find the perimeter of each figure.

10.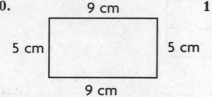
9 cm, 5 cm, 5 cm, 9 cm

11.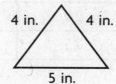
4 in., 4 in., 5 in.

12.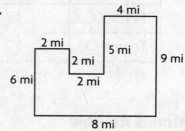
4 mi, 2 mi, 2 mi, 5 mi, 6 mi, 2 mi, 9 mi, 8 mi

_____   _____   _____

© Harcourt

Name _____

Patterns for Solid Figures

Vocabulary

Fill in the blank.

1. A _____ is a two-dimensional pattern of a three-dimensional figure.

Write the letter of the figure that is made with each net.

2.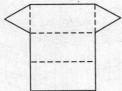

3.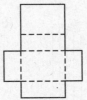

4.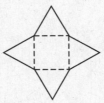

5.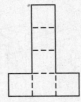

_____   _____   _____   _____

a.

b.

c.

d.

6. Which of the following nets would make a rectangular prism?

a.

b.

c.

d.

Mixed Review

7. $10\overline{)1{,}000}$

8. $14\overline{)0}$

9. $25\overline{)475}$

10. $32\overline{)256}$

11. Franz ate $1\frac{3}{8}$ granola bars. Aimee ate $2\frac{1}{8}$ granola bars. How many granola bars did Franz and Aimee eat in all?

_____

Name _____

Estimate and Find Volume of Prisms

Find the volume.

1.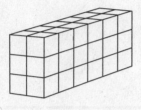

_____

2.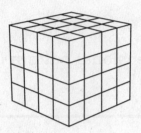

_____

3.

_____

4.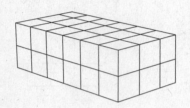

_____

5.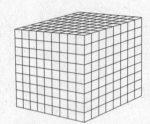

_____

6.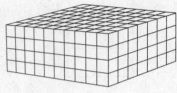

_____

7.

_____

8.

_____

9.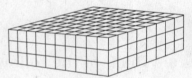

_____

Mixed Review

10. $\begin{array}{r} 17 \\ \times\ 6 \\ \hline \end{array}$

11. $\begin{array}{r} 247 \\ \times\ 48 \\ \hline \end{array}$

12. $\begin{array}{r} 89 \\ \times\ 17 \\ \hline \end{array}$

13. $\begin{array}{r} 478 \\ \times\ 45 \\ \hline \end{array}$

14. $\begin{array}{r} 112 \\ \times\ 39 \\ \hline \end{array}$

15. $\begin{array}{r} 222 \\ \times\ 31 \\ \hline \end{array}$

16. $\begin{array}{r} 52 \\ \times\ 44 \\ \hline \end{array}$

17. $\begin{array}{r} 63 \\ \times\ 12 \\ \hline \end{array}$

18. $\begin{array}{r} 678 \\ \times\ 18 \\ \hline \end{array}$

19. $\begin{array}{r} 456 \\ \times\ 48 \\ \hline \end{array}$

20. $\begin{array}{r} 75 \\ \times\ 36 \\ \hline \end{array}$

21. $\begin{array}{r} 67 \\ \times\ 58 \\ \hline \end{array}$

22. $\begin{array}{r} 159 \\ \times\ 43 \\ \hline \end{array}$

23. $\begin{array}{r} 517 \\ \times\ 62 \\ \hline \end{array}$

24. $\begin{array}{r} 802 \\ \times\ 24 \\ \hline \end{array}$

Name _____

Problem Solving Skill: Too Much/Too Little Information

Decide if the problem has *too much* or *too little information*.
Then solve the problem if possible.

1. There are 90 rocks in Joe's box. He has 45 different
   kinds of rocks in his box. The box is 12 inches
   long, 6 inches wide, and 4 inches high. What is the
   volume of the box of rocks?

   _____

2. Klamo likes to take pictures of animals in her
   backyard. She has over 100 pictures of animals.
   She keeps her pictures in a box that is 1 foot high.
   What is the volume of the box?

   _____

3. Spencer puts corn from his garden into wooden
   boxes. Each box contains 30 ears of corn. Each box
   is 2 meters long and 1 meter wide. What is the
   volume of the wooden box?

   _____

4. A cereal box weighs 1 pound. It is 12 inches high,
   6 inches long, and 2 inches wide. What is the
   volume of the cereal box?

   _____

Mixed Review

Find the area and perimeter of each.

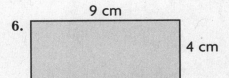

5. 16 ft / 9 ft

6. 9 cm / 4 cm

7. 18 mi / 23 mi

_____   _____   _____